# NIGHTINGALES IN BERLIN

# NIGHTINGALES IN BERLIN

## Searching for the Perfect Sound

## DAVID ROTHENBERG

*The University of Chicago Press*
*Chicago and London*

The University of Chicago Press, Chicago 60637
The University of Chicago Press, Ltd., London
© 2019 by David Rothenberg
Published 2019
Printed in the United States of America

28  27  26  25  24  23  22  21  20  19      1  2  3  4  5

ISBN-13: 978-0-226-46718-4 (cloth)
ISBN-13: 978-0-226-46721-4 (e-book)
DOI: https://doi.org/10.7208/chicago/9780226467214.001.0001

Library of Congress Cataloging-in-Publication Data
Names: Rothenberg, David, 1962– author.
Title: Nightingales in Berlin : searching for the perfect sound /
David Rothenberg.
Description: Chicago : The University of Chicago Press, 2019. |
Includes bibliographical references and index.
Identifiers: LCCN 2018052431 | ISBN 9780226467184 (cloth : alk.
paper) | ISBN 9780226467214 (e-book)
Subjects: LCSH: Music—Philosophy and aesthetics. | Birdsongs. |
Nightingale. | Nature sounds.
Classification: LCC ML3877.R67 2019 | DDC 781.1/7—dc23
LC record available at https://lccn.loc.gov/2018052431

♾ This paper meets the requirements of ANSI/NISO Z39.48–1992
(Permanence of Paper).

It is not becoming to humanity
That I should be silent
when birds chant praises.

—Saadi

# CONTENTS

# 1

## THIS BIRD IS RUINED FOR US

Are you surprised there are nightingales in Berlin? They have flown thousands of miles to get here, up from Africa and over the sea like refugees of the air. They sing from wells of silence, their voices piercing the urban noise. Each has his chosen perch to come back to each year. We know they will return, and yet when they do arrive every song still seems a wonder.

Of all the days to schedule a midnight concert in Berlin's Treptower Park, we have somehow chosen May 9, the one night people descend upon this park in the thousands. It is the sixty-ninth anniversary of the end of World War II. The park will be full of people when the birds begin to sing. The location itself lends the timing further significance. This is where the great Battle of Berlin is remembered, during which a hundred thousand died in less than two months. Here stands an extravagant war memorial, built by the Soviets to commemorate their victory in what was once East Germany.

Upon entering the memorial grounds, one crosses a jagged abstract Constructivist gate with a menacing hammer and sickle. At the far end, about five hundred feet away, is a ninety-eight-foot-tall bronze Russian soldier in a long war coat holding up a child, as if to reassure the boy that he is safe from all the horrors commemorated around him. Beneath the towering statue are sixteen heavy concrete sarcophagi with realist murals carved into their surfaces, depicting the course of the battle and the courage of its commanders, including more than one image of Stalin himself.

Newly united Germans restored the monument as part of a Soviet agreement, but the explanatory text at the entrance indicates that they seemed embarrassed by the whole thing: "Although the grandiosity of this monument might seem inappropriate to current memorial style, at the time the language of commemoration was quite different. The war memorial at Treptower Park should be considered one of the finest examples of Soviet Socialist Realism and has been restored to the best possible level of authenticity."

Although the weight of history bears down heavily here, it is surrounded by quiet forests, a lake, and a beautiful riding path on the shores of the Spree River. It is the most graceful of any of the city's parks, with its mix of plantings, *grandes allées*, and crumbling vestiges of Communism. And it is here that a few dozen male nightingales establish their territory every spring, and we wander in the dark shadows of this concrete history to engage with the most ancient music in the world.

Berlin is the best city in Europe to hear the song of the nightingale, and the right time to hear it is from late April through late May. This is when the male birds return from their migration to Africa to establish their territories, sing for their mates, and nest together with them to raise their young. By early June the song thins out; the birds remain in the trees until August but become much quieter. As the evenings cool once again, they head south, not to be seen until the following year, when they will come back on schedule, often to the exact roosts they established the year before.

Nightingales are connoisseurs of sound. Our human clamor doesn't seem to bother them. In fact, they might like the challenge of our noise. Of all songbirds, nightingales are the two species, *Luscinia megarhynchos* and *Luscinia luscinia*, most inclined to sing in darkness as opposed to early morning light. As such, they underscore all those human romances and yearnings of the clandestine, indecorous dark.

These birds are celebrated in myth, song, poem, and story, and I for one had read much about them before I ever heard one. The poet Matthew Arnold, hearing the nightingale as an ancient and omniscient traveler, wrote in 1853:

O wanderer from a Grecian shore,
Still, after many years, in distant lands,
Still nourishing in thy bewilder'd brain
That wild, unquench'd, deep-sunken old-world pain ...

Arnold heard a shade from an ancient myth before he could admit this was a real bird. Most of us feel the same when we hear our first nightingale. When I finally did encounter my first real one, I could not believe what I was hearing. This song was *weird*. A series of detached phrases. A mix of rhythmic chirps, spread-out whistles, and funky contrasting noises. It was neither mellifluous nor melodic, like the heavily praised tunes of the hermit thrush in North America or the blackbird in Europe. This was, rather, an unusual rhythmic assault. I had no doubt that it was music, but a foreign music, another species' groove, a challenge for humans to find a way into. I wanted to *know* his method, and began to imagine some way to one day join in.

So can we take it seriously as music? Transcription into notes and measures does little justice to the song of the nightingale. Sonograms can help, but those images come across as a secret scientific code. Johann Matthäus Bechstein, a forester and pioneer conservationist in Germany, attempted to transcribe the nightingale song into words in 1795 in his *Natural History of Cage Birds*:

Tiuu tiuu tiuu tiuu,
Spe tui squa,
Tio tio tio tio tio tio tio tix
Qutio qutio qutio qutio
Zquo zquo zquo zquo
Tzü tzü tzü tzü tzü tzü tzü tzü tzü tzi
Quorror tiu zqua pipiqui.
Zozozozozozozozozozozozozo Zirrhading!
Tsisisi tsisisisisisisisi
Zorre zorre zorre zorre hi;
Tzatn tzatn tzatn tzatn tzatn tzatn tzatn zi,
Dlo dlo dlo dlo dlo dlo dlo dlo dlo dlo
Quiro tr *rrrrrrrr* itz

Lü lü lü lü ly ly ly ly li li li li
Quio didl li lulyli....

It does not read or sound like anything human. The real nightingale is nowhere near as melodious as he is often described. The literary scholar John Elder was as surprised as I on his first trip to Europe to hear said bird and concluded that our love of the nightingale's song has as much to do with the range, energy, and ability of the song to carry far through the trees as with anything musical in the song. There is such passion in the tones that it seems as if the bird would die from his music if he had to, as those ancient myths want to imply.

I make music with the sounds of other species and seek them out all over the world, playing my clarinets along with them, living in their habitats, to create music with sounds I might never understand, sounds not sung for me. I try to change my own tones so that between clarinet and nightingale we might produce a sound that neither of us could make on his own.

The fact of being able to do this in Europe's second-largest city, a burg of nearly three million people, gives me a special kind of hope. Even though nightingales no longer sing in Berkeley Square, as a famous London song once intoned, they are everywhere in Treptower Park, this green riverfront oasis where East and West Germany once lay divided.

It's not only in the parks that one comes across nightingales in Berlin: some prefer trees in quiet urban neighborhoods, behind a playground, or in an abandoned lot, where their tones may be enhanced by the amphitheater effect of surrounding buildings. There is one famous bird who alights every night atop a traffic light at the main junction in Alte Treptow, adjacent to the S-Bahn station and the entrance to the park, as if he has specifically chosen the noisiest possible spot in order to *prove* that his sound can be greater and more tireless than any noise around him.

Today Berlin is an international city where those aching to make culture find a place to call home. You can become a part of any number of scenes or create your own—there is always a new, not-yet-hip neighborhood ready to be colonized by the next group who dares to overhaul a burned-out building or fix up a crumbling factory. Berlin is still the cheapest capital in which to set up shop in Europe. It's a place where people innovate and don't demand to be paid for it. One needn't work two frenzied jobs to pay for the privilege of creating culture, as one might in New York. The city makes the music for you. Maybe the nightingales think that way as well. They too are outliers, with one of the strangest and most complex songs of any bird on the planet. Their song has a definite style and aesthetic, one we humans can't easily pinpoint. H. E. Bates understood this many decades ago:

> The song of the nightingale has some kind of electric, suspended quality that has a far deeper beauty than its sweetness. It is a performance made up, very often, more of silence than of utterance. The very silences have a kind of passion in them, a sense of breathlessness and restraint, of restraint about to be magically broken. It can be curiously seductive and maddening, the song beginning very often by a sudden low chucking, a kind of plucking of strings, a sort of tuning up, then flaring out in a moment into a crescendo of fire and honey and then, abruptly, cut off again in the very middle of the phrase. And then comes that long, suspended wait for the phrase to be taken up again, the breathless hushed interval that is so beautiful.[1]

Berlin is home not only to the most urban nightingales in Europe, but also to the most nightingale scientists. They work out of a lab at the Free University in Dahlem founded by Dietmar Todt, now retired. Today the lab is run by Constance Scharff and houses one of the foremost neuroscience research groups in the world. Within it Silke Kipper runs a multi-year study of the nightingales of Treptower Park.

What intrigues these scientists is that nightingales acquire their music; they are not born with songs hardwired into their brains. In the animal world only whales, dolphins, songbirds, and humans can learn through sound. Not chimpanzees or other primates. Not wolves, dogs, or cats. And, most important for science, not rats, which are the animals most analyzed and understood.

Science wants to know how animals evolved what they call "vocal learning." This is most easily studied in birds, and biologists have chosen the Australian zebra finch as a model species for the study of this phenomenon. Thousands of scientists around the world study the brain and song-learning abilities of this colorful bird. Zebra finches have a very simple song. Simple in structure, that is, but not agency. How it is produced and appraised is complicated enough to keep legions of scientists busy for years.

Enter the nightingales. Their song is as different from the zebra finch's as one could imagine. It is loud, long, intricate, structured, and musical—an extreme example of what evolution can produce through sexual selection, as generations of female birds have preferred increasingly refined and nuanced songs. How has this refinement progressed? Is it a matter of balancing noise and tone, whistle and crack, similarity and difference, an æsthetic as elusive as any human style of music? Depends how much you know. What we know depends on the questions we ask.


It's after 11:00 P.M., and the humans are slowly filing out of Treptower Park now that the annual memorial concert is over. I'm wandering around the park, hearing the nightingales tentatively begin to sing. I stop for a beer at a small kiosk, and a guy bumps into me and hears me speaking English. "Hey, you are American? What you doing here? On this night of all possible nights!?" He glares at me from a few inches away, vodka on his breath.

His friend pulls him back. "You must excuse Yuri," says his companion in a heavy Russian accent. "He has had a bit too much to drink."

Yuri spits and rumbles away, staring me down as he turns. His friend is more accommodating. "My name is Oleg. May I ask you a question?"

I take a slow sip of beer. "Sure. Why not?"

"Why do you Americans say you won the war? You lost 25,000 men. Russia lost 25 *million*. It was not your war to win."

My history was hazy: Weren't we and the Russians on the same side in that war? Far more Russians did indeed die, however. After all, this was their continent. And here we are, drinking together where one of the war's bloodiest battles went down. The fields are green and the trees grow tall.

My thoughts return to nightingales. The BBC recorded Beatrice Harrison playing Elgar and Brahms on her cello to the nightingales in her garden in Kent each spring. It was the first outdoor radio broadcast ever when it was first tried in the 1920s, and the ritual was repeated every year thereafter.

Until the war. Just then, as they started to record the birds, the roar of Allied bombers was heard, and the station went silent so as not to alert the enemy. Only years later was the haunting recording of Royal Air Force bombers humming along with the singing nightingales released, a solemn reproach that nature's music will not be quashed by our need to fight and kill.

These enigmatic songs are in our midst, ever just beyond our power to comprehend. I'm sure they were trying to sing throughout that fateful spring of 1945 even as so many Russian lives were being lost in their attack on Berlin.

Nightingales sing through all wars. One soldier in World War I also heard the most beautiful treetop tunes in battle. In one of the classic books on our bird, *The Nightingale: Its Story and Song*, Oliver Pike writes that one of the best performances he ever heard from a *rossignol* was in the middle of a battle in a French forest in 1916:

> The wood was lit up with vivid flashes, while overhead a score of star shells floated, flickered a moment or two, and died down. As the time wore on the shells increased in violence, the whole ground seemed to be trembling with the force of the explosions, while suddenly there broke out a glorious melody.
>
> At first the nightingale seemed doubtful, and there were pauses between his bursts of song, but as the bombardment increased he took up the challenge, and if we had searched the world over, it would have been difficult to find a greater contrast between the beautiful harmony of his song and the awful discord of the bursting shells. But as suddenly as the song began it ceased, for a shell burst under the singer, and the tree in which it was perched was blown

to matchwood, and the small bird which had entertained the waiting soldiers was killed together with five brave men who were near.[2]

Birdsong in battle stands starkly between beauty and terror.
Pike goes on to give practical advice, as far back in 1932, to those
who would incite the songs of nightingales:

> Time after time I have proved that if you want to get the best efforts
> out of a nightingale you *must* provide an opposition entertain-
> ment that will almost drown its song. The raucous noise of a klaxon
> motor-horn will often start a bird singing. I suggest that the next
> time the BBC attempt to broadcast the song of the nightingale they
> should provide a battery of big drums within a hundred yards of
> the singer, then listeners will hear what wonderful music this bird
> is capable of giving.[3]

After almost an hour discussing the weight of history with Oleg
and Yuri, we come to some kind of agreement, if only because I
agreed to listen. "Well" (Oleg puts his arm around my shoulder
as he tries to steady himself), "at least there is one American here
I can trust," he says, before he and his friends wobble off into the
night.

Everyone seems to be deserting the park. I can't believe it. It's
11:30 P.M. and the festivities are over. That's just about when Berlin is supposed to wake up! At least the nightingales are waking up.
At midnight I will meet my audience and we will all head into the
night, seeking the perfect moment in a nightingale song in which
there is still room for humans to enter.

A small group of dedicated interspecies musical adventurers arrive at the S-Bahn station at midnight. We are not afraid of rain,
or of the few remaining Russian revelers. And we know that nothing scares nightingales. Rosa Luxemburg once noted this, sitting
by her prison window:

> At six o'clock, as usual, I was locked up. I sat gloomily by the win-
> dow with a dull sense of oppression in the head, for the weather was

sultry. Looking upward I could see at a dizzy height the swallows flying gaily to and fro against a background formed of white, fleecy clouds in a pastel-blue sky; their pointed wings seemed to cut the air like scissors.

Soon the heavens were overcast, everything became blurred; there was a storm with torrents of rain, and two loud peals of thunder which shook the whole place. I shall never forget what followed. The storm had passed on; the sky had turned a thick monotonous grey; a pale, dull, spectral twilight suddenly diffused itself over the landscape, so that it seemed as if the whole prospect were under a thick grey veil. A gentle rain was falling steadily upon the leaves; sheet lightning flamed at brief intervals, tinting the leaden grey with flashes of purple, while the distant thunder could still be heard rumbling like the declining waves of a heavy sea. Then, quite abruptly, the nightingale began to sing in the sycamore in front of my window.

Despite the rain, the lightning and the thunder, the notes rang out as clear as a bell. The bird sang as if intoxicated, as if possessed, as if wishing to drown the thunder, to illuminate the twilight. Never have I heard anything so lovely. On the background of the alternately leaden and lurid sky, the song seemed to show like shafts of silver. It was so mysterious, so incredibly beautiful, that involuntarily I murmured the last verse of Goethe's poem, "Oh, wert thou here!"[4]

Why so much song from one little brown bird? It is indeed excessive—and risky. One male nightingale singing on one perch for hours on end through the night could easily be picked off by a tawny owl. He takes the chance. According to Darwin's theory of sexual selection, this bird has honed such beauty only through the connoisseurship of the female nightingale. She alone knows just what kind of song is the best song. The evolutionist Richard Prum says this is why the nightingale's music evolves in an "art-world," only making sense together with the evolution of an aesthetic of appreciation among female birds.[5] We humans can listen, study, surmise, calculate, measure, and dare to join in, but the full inhabitation of the nightingale aesthetic continues to elude us. We are not yet inside the nightingale's own genius. But that doesn't stop us from trying.

Do nightingales *like* making music with people? The most rigorous study of nightingale response to playbacks of their own species' songs, conducted in Berlin in the 1970s by Henrike Hultsch and Dietmar Todt, discovered three ways a nightingale may respond to a strange new music in his midst. First, if he feels his territory is threatened, he will try to interrupt the unfamiliar sound—what the scientists called "jamming the signal"—thereby preventing any foreign message from coming through by getting in the way of it as much as possible. That's the aggressive response. But he may respond differently. A male nightingale who is confident in his territory, who doesn't consider you and your clarinet or iPad or voice or cello a threat, will listen to what you play, wait a moment, then respond with his own short song, and then pause again. If you give him some space, play a short phrase, and stop, the whole exchange is considered a friendly acknowledgment, with each musician trading ideas, leaving space for the next, accepting that we each have our place and our song.

Third, a nightingale who considers himself at the top of his game—the boss bird, the best singer in the whole park—will do whatever he wants, maybe interrupting, maybe leaving space, singing however long it pleases him, because you matter not in the least to him, convinced as he is of his greatness. He sings as if no one is there but himself.

We've all met musicians who fit into these three categories.

From a musical point of view, distinguishing between interruption and sharing could get quite blurry. What one person hears as jamming the signal could, to another, come across as just plain jamming, trying to make interesting music together. This is because music is far from a simple sign. It depends on what one believes music, in either a human or an avian context, to be all about. Perhaps artistry and form constitute not just an advertisement of territory and skill, but an attempt to work together to create something no one species could make on its own.

It was with this idea in mind that I felt compelled to bring people and nightingales together to make interspecies music in the first place. Through flash messaging and the wiles of social media, somehow at least a hundred people had gathered at the Treptower Park S-Bahn stop by midnight to follow us to the ideal location, one copse away from the river's edge, where our favorite bird, with whom we had practiced on earlier days, was ready for showtime.

I am ready to play clarinet live with the birds, my first time for an audience of more than one. Playing along with a nightingale becomes a direct window into the unknown, a touch of communication with a being who doesn't share our language. The game of pure tones jarring against click and buzz becomes not a code but a groove, an amphitheater of rhythms in which we strive to find a place.

The birds leave space for each other; they are in that back-and-forth state, standing their ground, thus welcoming me perhaps more than usual. Even the occasional human cry in the distance has its place: all sounds are welcome. Finally a screech. Is it someone blowing against a blade of grass? Will that silence our bird? Absolutely not, nothing will. For he is born to sing.

I want to convey to you something special about jamming with another species, but I don't know if jamming is the best word. Does that suggest something frivolous to you? Musicking? Playing along with? Finding common ground? Interspecies music, of course, is music that no one species could make on its own. And the whole, if it works, should be greater than the sum of its parts, just as nature is greater than any one species in its midst. We all have our place, and no species is an island. We enhance ourselves by paying more attention to the rest of life.

One song or many: what is that bird up to? Many songs in a row, up to a few hundred in a song "bout," or one multiplicitous song out of many riffs or phrases? How much space between the riffs? How much *listening* goes on in those silences? I want to listen as much as the bird does. We don't fight each other for attention—we strive for mutual comprehension. The music we make together is more than a war.

People always ask me what it feels like, and my answer is never good enough. All I can do is play music attuned to the moment and the presence of the birds, leaving space for their songs and their silences. Treat them as equals with whom I cannot speak. It was uniquely moving to bring a patient audience out into Treptower Park an hour after the Russian victory festivities had subsided and a strange calm descended on the night. Only then did the birds comply, as if they had enjoyed all that noise and human celebration of the war's end.

They are not afraid of us. They coexist with us, hiding in their nettle fortresses, waiting for the right moment to sing. We honor their sound by calling it a song, by deciding it to be something worth taking seriously as music and finding a way to join in. I say this again and again, a refrain in and of itself. The same simple message, one easy way to make nature matter. Listen to it. Don't sit passively, but love it enough to want to play along. It's got room for you.

This nightingale is one famous bird. Every language has something clever to say about him, trying in vain to capture a sound not made for us to understand. Nothing can stop us wanting to make sense of it. In some tongues, his name means "a thousand voices," in others "the sound of the night." *Eos, solovej, fülemüle, urretxindor, ushag-oie, passirillanti, rietumu lakstigala, satakieli* and *bülbül*, beyond the more familiar *rossignol, Nachtigall, ruiseñor*. Some seem unparsable, strange onomatopoetics that mirror his beguile. I imagine in one of those languages the word for nightingale actually means "rhythmic madman," and I keep thinking that the rhythms matter more than the notes for this avian singer, for the spaces between the beats are essential for the possibility of our collaboration. The bird leaves room, for his peers or anyone. He taunts us with the possibility to answer.

Like any foreign music, the nightingale's tones are more accessible the more time you spend within them. When you think you've got it, keep going. Listen further. Don't tune out once you have identified the species behind the song. Remember that hearing can be forgetting the name of the thing one hears. Know how much more the song means to the bird than it will ever mean to you. As a musician, dare to join in.

The nightingale pauses in his renditions after each phrase to give you a moment to reflect; to challenge, acknowledge, or ignore, depending on your mood. It is a microcosm of all music, a study in similarity and difference, in repetition and novelty, in noise and rhythm versus melody and pure tone. It is always more and less than anything we can add to or take away from it. Its rhythms are not boring, its melodies continue to surprise. We can never quite *get* it, this song intended for nonhuman ears.

Just as I prepare to get my instrument out, we see them—our friends, the scientists, Silke Kipper and her associate Sarah Kiefer, leaders of the Free University's nightingale research project. They have chosen this exact moment to do some playback experiments with this very bird. They are not pleased to see us. "What are you doing here, David? You know this is our study area. We don't want you ruining our data collection."

We had talked about this earlier. "I know," I say, with apology. "But this bird is so special. We've listened to many, but we keep coming back to him."

"How do you know?"

"I was playing here just the other night."

"Playing what?

"Clarinet. Voice. Some electronics."

"What kind of electronics? I heard you just now, and it sounds like you have nightingale songs on that iPad."

"Yes," I confess, "we were sampling the bird and playing him back his own song. Looped. Remixed. Pitch-changed. Sliced and diced."

The glow of my iPad illuminates a sense of betrayal on her face. "This bird is *ruined* for us!"

"What do you mean?"

"We don't care if you play clarinet or cello or sing to him, but playing him back *his own song*! That is a playback experiment, that is what *we* are trying to do. I hope you have the proper *permits* for conducting experiments with a wild animal!"

"We are just making music together with birds."

"You have compromised our research subject. Messing with his brain, his whole sense of aesthetics. Who *knows* what your music has done to him!"

I am a bit surprised by her anger. "This isn't exactly a pristine wilderness, is it? A few hours ago this place was flooded with Russian songs celebrating the end of World War II! Did that ruin the nightingales? These birds hear all kinds of human sounds, all day, every day."

"Sure, but it is their *own sounds* that they care for the most."

"How do you know that?"

"Humans care most about other human sounds. That's how we talk to each other or sing together."

"Well, some of us like to sing with animals."

"And have no idea whether they like it or not."

"I can hardly tell whether other *people* like my music. But I learn from their response, as I learn from the birds."

"What if you are upsetting the birds?"

"Nothing we ever do seems to make them stop singing."

"If you don't chart their mating success you won't know whether you have impacted their ability to attract a mate and procreate."

"I don't know about that with humans, either, but we still spend an awful lot of time making music." At this point, the small crowd listening in on our conversation snickers a bit.

Silke sighs. "You win. Look, I see you have a lot of people gathered here. You don't want to disappoint them." She turns, dejected, mumbling to herself. *Ruined, ruined, another experiment, ruined …*

"Wait!" I run to stop her. "You're completely right, we shouldn't be here, I know this is your research area. From now on we'll stick to other parks. So much has been learned from your lab, and we wouldn't want to mess it up. We're going to tell everyone to move down the river to another bird, at the edge of the park. Carry on!"

And so I make the whole crowd get up and keep walking to the next bird, and a new sense of sound. I really don't want to ruin someone else's data. I want the scientists on my side! I will go on to write papers with some of them. I want to encourage them to quantify musicality, because I don't want to do that myself. The rigor of science is important, and someone should be applying it to the beauty of nightingale songs. More and louder are not necessarily better, despite being easier to measure.

The next bird does not disappoint, either. He is ready to defend his line against a hundred eager humans, none of whom pose any real threat but who sit reverently before him as the inscrutable song begins. I imagine myself—half man, half bird—strangely poised between nature and technology, darkness and light, dirt and sky. I breathe in and blow out a sound that reaches toward that million-year-old song.

The nightingale keeps singing his inscrutable melodies. I suppose I *am* interrupting him, as the scientists say, but as a musician I feel he is enjoying being taken seriously by another musician of another species. In listening to this one bird, we listen to all. We sift these interactive songs together with the silence. All the world's sounds layer on top of one another as we try to remember everything we have ever heard. I think to myself: I'm going to figure this out. I'll learn what to tell the scientists so they can accept the idea that nature and humanity can fit together. I don't want to ruin their data, but neither do I want to deny that music might help us make better sense of the environment around us. I will work to convince you all of these things.

I begin by playing together with this fabulous bird, knowing nearly nothing. One year later I will be back, adding a few more musicians to this interspecies mix. I will read and listen deeply into this topic, trying to learn as much as I can. Maybe it will matter, maybe not. Now I know a few birds, and I know the trees to which they will return. Let us see next year if we are right.

# 2

## THE SHARAWAJI EFFECT

When a nightingale sings, some of us hear poetry, while others see numbers and charts. How will I know when his song "fits" into the place where it's sung? I will need to listen outwardly for that. These nightingales we have found are not just anywhere, but in Berlin. I believe they sonically fit into this place and uniquely make it sound alive. And that hunch leads me to something called the sharawaji effect, which first I learned of from a Swedish expert on singing crickets.

Lars Fredriksson looks more like a Chinese holy man than most anyone you would run across in China, except perhaps for his piercing blue eyes. He wears a long tattered robe, sports a graying Fu Manchu moustache, and speaks fluent Mandarin. He would probably would look out of place anywhere. He prefers to go by "Mr. Fung."

Mr. Fung spent many years raising 108 crickets in his tiny apartment in Stockholm, trying to teach all those different species to finally get along and sing together—a story I told in detail in my last book, *Bug Music*. His wife had gotten sick of all the noise, and his employer, the Royal Swedish Library, had closed its doors on its Chinese collection, where Fredriksson was the curator. He was preparing to leave town for the Far East yet again, this time to find his version of the perfect sound. He knew just where he would go.

"There is a pavilion atop a hill"—and here he smiles, looking me straight in the eye—"miles from nowhere far in the East. The country is irrelevant. I don't want to tell you. It is three days' walk from the nearest recognizable town. You won't find this place on Google Maps." We are speaking in a beer garden on Södermalm. Lars con-

tinues: "Think of a lazy, humid late evening that hasn't quite decided for itself whether it is summer or fall. The distant mountains
are indistinct; their color is fading from greenish blue to gray and
then black. Doesn't really matter what they look like because you
can hardly see them—the sky is so heavy with moisture that you
wish it would rain, but it's not going to anytime soon."

"So it's perfectly quiet, pristine?"

"Absolutely not. The cicadas are screaming, the crickets are fid-
dling, thousands of layers of unraveling rhythms are going on as
they have for thousands of uneven years. It is a complete and en-
veloping sound. You might think it is just like the sound of a warm
August evening anywhere in the world, but it is not. I would venture to say this is the most beautiful sound in the world."

"I want to hear it."

"I'm not sure you do."

"Why not?"

"Our love of sound is something subjective. *I'm* the one who
finds this to be the most beautiful sound in the world. But I don't
want to say it is most beautiful only to me. My feeling of it is absolute, total. It completely satisfies. But I know this is only my satisfaction we are talking about. It changes nothing, does absolutely
nothing to help anyone else on this miserable decaying planet."

"I'm confused. Do you love this sound or does it leave you longing for more?"

"There is no love for nature without longing."

"I would like to visit you in this place."

"You can't. And you shouldn't really want to. It will very likely
mean nothing at all to you, and certainly not what it means to me.
Sound and love are the most personal of attributes. Inside a person, and out there in the world." He pauses to sip his beer. "Ever
hear of the sharawaji effect?"

"No."

"It is the most elusive of all sound effects. And the most important and beautiful. Many people do not think it exists at all."

"Tell me more."

"It is a rare and ancient name for the absolutely perfect sound."

Even the etymology of the word is woefully obscure. Mr. Fung tells me that in 1691 Sir William Temple wrote, "The Chinese have a particular word to express it, the *beauty of studied irregularity*, and where they find it hit their Eye at first Sight, they say the Sharawaji is fine."

Sharawaji? Doesn't sound like a Chinese word at all. Japanese, then? Just seventy years ago E. V. Gatenby said that the word sounded like an archaic form of the Japanese *sorowanai deshō*, what you say when the two parts of some design do not match. The archaic form, *sorowaji*, died out four hundred years ago.

"Lars, that word hardly seems real."

"Right you are. Imagine a Dutchman sailing the Far Eastern seas and trying to make sense of the word. *Sharawaji*. Ergo, there you are. It sounds immediately more international. Nearly Persian in its universality. It comes from everywhere. And nowhere. I think I've found it already."

"Where?"

"In dreams, perhaps—the best of dreams. You can't look for it, but have to listen." He paused for a moment and closed his eyes. "I start to hear glimmers of it more and more. The potential for its appearance is right around us, but few of us take the time to learn. In my concrete apartment block, the ventilation holes give a clear tone to the sound of the wind. Wind is otherwise very difficult to record, since it is air moving fast against the world. Against a microphone it just produces distorted noise. Against trees you can definitely get something. The wind, on its own, makes no sound at all. It must move against the world for its true sound to be heard. And this is only one part of it. The sharawaji effect ought to unite our own song with the wind, with our place, with our touch, with a sound that knows exactly where it is in the world."

"I'm not sure I follow what you're talking about."

"You'll never follow it until you hear it." Lars shakes his head. "You must leave behind your expectations and commence your journey. Or find the journey in the places you have already gone, in the sounds that define the arc of your whole life until now. Only then will you gain the skills necessary in order to hear the future."

I once heard the computer scientist Marvin Minsky say that there is far more music in the world than there ought to be. This was at a conference, held at Canada's Banff Centre in 1993, on music and the mind in relation to the future and the past.[1]

Minsky was saying that humans spend much more time on music than we ought to, that although it must really be somehow essential or necessary for human life and survival, no one has come up with a good biological reason for why we spend so much time listening or thinking about it when it seems so useless in the grand scheme of things. That got me thinking how odd it is that I don't really like all that much music.

I reduce whole albums to one song or even less than a song. I edit down pieces that are more than ten minutes long because there's no way I'll ever want to listen to one thing for that long; my attention will wander, I'll get bored and wish I were doing something else. I am playing my simple instrument into ever more confusing software programs, trying to turn the sound into something more than it is by means of trivial enhancements, but they are all just sound effects and there is only so much they can accomplish. Still, I ask for more of them and try new things out, as do many musicians today, because every day new effects are being released. They are instantly accessible, and we can't wait to try them because we are always hoping the next one will overshadow all predecessors and will transform the way we listen, think, or play, transporting us to some new, more real, greater, better world than the present one in which we are stuck—a world where sound throws itself out in the air and just sits, lost, stares back, appears and is gone, a sound that was and now isn't and will never be again. I know that comes across as hopeless, but I guess if I were completely satisfied with anything I played I'd be like Artie Shaw, who, after recording with his Gramercy Five in 1954, said, "Well, that's perfection," and put down his clarinet, never to pick it up again for the fifty remaining years of his long, crazy life.

If the sound is never quite right, you need to keep playing and hope you stay unsatisfied for years. There's no need to be at peace if you want to create. Before listening to single birds, I started wondering about all those times I had been asked to find a perfect sound, and whether or not I had ever succeeded. Had I ever really experienced the sharawaji effect?


Andrea Galvani is asking me for the sound of a collapsing iceberg. Again. He doesn't seem to remember he asks me for such a sound every once in a while. Whenever I do send him something, the sound is never quite right. Possibly this is because the actual sound of a collapsing iceberg is a giant muffled rumble that sounds a bit too much like a distant subway train. Galvani has *ideas* about what this sound should be: a huge, cascading, crumbling crash of solid against liquid. I tell him that's what sound design is for; we make the sound you heard in your dreams. How we do it must remain a secret. Maybe I'll even use a subway train as the source just to confuse him. Maybe I will just record myself talking in my sleep while I too am dreaming of icebergs collapsing. Who isn't dreaming of the icecaps melting these days as the warming atmosphere produces little more than bad news? We all know the problem is so vast that no matter how we change our own ways, the ice continues to melt, the great frozen wastes break apart with sounds that are never as vast as we think they should be. In the beginning was the sound, in the end comes the sound. In our imaginations we articulate that sound and turn it into art.

Galvani is an Italian photographer who places incongruous objects in the middle of otherwise wild places. He once released a herd of black-and-white rabbits onto a Swiss glacier and photographed them hopping all over the snow. Half of them were nearly impossible to see, but all except one were recovered. The photo was blown up to impressive artistic size and displayed in a gallery in Milan. He and I once traveled around the arctic archipelago of Spitsbergen on a red schooner full of artists, and on the trip he photographed me wearing a long Muji coat, standing up in a Zodiac raft playing a gleaming soprano saxophone against the steaming wall of a melting glacier.

A wire descended from the saxophone, suggesting that the sound was being broadcast underwater and played along with singing whales. I have engaged in such an activity many times, although

FIG. 1. David Rothenberg playing to imaginary whales in Svalbard

usually in Hawaii, where there is far more music underwater and the conditions are more amenable to performance. Those musical adventures were fun but look nowhere near as poetic as Galvani's picture of me balancing in the rubber raft playing alone to arctic whales at 78 degrees north latitude, only 350 yards from the North Pole. Still, I use that picture all over the Internet, as it gives me a beautiful, stark image of what it is I think I am doing.

I have Galvani to thank for this emblematic image, and the least I owe him is the sound of an iceberg collapsing. Yet even when I listen to recordings of such grave contemporary moments, they do not do the conceptual job this sound needs to do. Just as Australian kookaburras (*ooo ooo ooo ah ah ah ah ooo ooo ah ah*) have been used for decades as a cinematic stand-in for screaming monkeys, the sound I create for this disintegrating glacier has to stand for the very possible end of ice here on Earth, taking human life away with it.

Images of collapsing glaciers are now ubiquitous, as the looming reality of global warming weighs heavily upon any image we conjure up of the future. The superb film *Chasing Ice* chronicles the efforts of James Balog and his team of nature photographers to depict the drawn-out recession of great rivers of ice all over the world. Still, the grand thunder of a collapsing glacier is the most graphic image in the film. Balog sent two young assistants to wait for an entire month on an isolated Greenland hilltop overlooking a large glacier, a huge piece of which was bound to fall off sometime over the summer. But documenting nature takes time, as all those David Attenborough outtakes so gamely show. Balog's crew could only wait. And the film beautifully shows how they almost miss the moment, how tired they had grown of waiting. But then it comes, a cloud of white dust, an unfurling plume, and a great rumble fills our ears. It is a wide, abstract sound, nothing that can easily be contained. We need the whole backstory to love it.

Calving, it's called, when a glacier casts off a piece of itself. To form an iceberg, to beget the next generation. Or just, just as easily, to melt away into the surrounding water, or vaporize into the air above.

Do we need to know the sound is real to trust that this tragedy is befalling our planet? Certainly not. The drama of sound is always being reinvented, massaged into being out of abstract possibilities. Galvani has asked me to give him a sound, and I will give him a sound. But I will *not* tell him how I made the sound. I don't even know how I will make the sound. I will just close my eyes and imagine how I want it to sound. I think of the imaginary helicopter made up by Walter Murch in the signature opening moments of *Apocalypse Now*, created using only synthesizers and no real rotor blades at all, along with the sound-art collage of the cassette tape Kurtz has sent downriver for Willard to sample before he heads into the jungle to confront this madman. Murch is still doing it years later when he concocts the perfect *bleep* as the Large Hadron Collider is turned on in the 2013 film *Particle Fever*, about the ultimately successful search for the Higgs boson particle.

I'm still imagining how to present the significance of breaking ice while not knowing exactly what Galvani wants or expects. He's always aiming to shock with a smile, coating motorcycles in mud, filming a hundred glowing rabbit eyes on a ski slope in darkness, setting off smoke bombs in the mountains. I am inclined to help him get his sound, though I know he hasn't a clue about what he wants. I read today about William Gaddis's final work, a history of the player piano, in which he rants about how an artist can easily be done in by the collective weight of his own images—let the art itself, not your life, exhaust you, and hope there is enough in it to matter all that much. To nearly do in its creator, if not anyone else who sees or hears ...

We fear the disembodied sound. We always want a voice to sound as though it is situated somewhere, real or imagined. Evan Eisenberg nailed this in the best book ever written on listening to records, *The Recording Angel*: "When a record is fitted over the platter, a transparency is fitted over a segment of space and time. The effect is a double exposure. If the music is worth its salt, it will assert itself as the true reality, and all the lovely furniture of one's room will seem a mere picture, a veil of Maya."[2]

There are enough people like me who prefer when the illusion is enhanced, when the sound comes to us in a room that could never exist, with some parameter extended beyond belief into the artificial, or the hyperreal. Behold: the supersound! More than the big bang and less than the zero-sum game. It's as hard to write about music as to contain the shadow of winter heat smoke falling on the terra-cotta roof tiles up here on the sixth floor on this clearest day of the year I have begun to spend here in Berlin.

The imaginary iceberg, invented in a city. A music made out of opposite kinds of sounds crashing together to make something powerfully new and strange: I am seeking sharawaji. When a sound like an airplane taking off makes its way into the beginning of a Joe Zawinul groove, or when Peter Gabriel begins "Sledgehammer" with a now-dated 12-bit sample of a shakuhachi on a Fairlight, I still like it and grin: sharawaji. When I wander at night with the headphones cranked up trying to record the perfect natural blend of katydids—*chh chh chh ing*—in the night at just the right moment when the sound balances, I lie in the midst of sharawaji.

It's a strange word, but I aim to make it familiar through these stories, which trust in the ultimate power of sound. Bang, whimper, everything in between, searching with the ears for something that is almost never there. No one said finding the sharawaji effect was going to be easy. Perhaps no music can contain it. Maybe it's nothing more than the wind heard whistling through an airy alpine cabin, or the fact that a bird would choose to sing along with the rickety clack of a midnight train.

The sound artist Peter Cusack has traveled the world asking people: *What is your favorite sound?* and thereby leading so many to wonder how we decide what noises we like, what tones matter. The sharawaji is the sound that makes us speechless, a music found to be so beautiful that no one could have composed it but by the confluence of noises only circumstance makes meet.

Sharawaji tells you why you should forget, in order to love and bathe in the most beautiful sounds you never knew existed. The coyote howls with the Doppler-shifting bus, the rock concert melds with the pile driver in the street. The luminous echoes of the aeolian harp are heard, too, in the mundane vibrations of telegraph wires. The best sound on the radio is the static between one station and the next, when the music doesn't pretend to be clear, when it knows itself to be a tangle of waves traveling through the air.

Sure, you may find your sharawaji where it finds you, but all artists want to believe they can make up something that has such an effect, by subtly blending one sound with another, by combining simple effects into something more complicated, by never being satisfied.

I don't want to extract something out of sound that just isn't there.

Galvani has written back to me. He has more ideas on the glacial sound. Apparently his plan is to imagine that the sound he hears from the iceberg could be loud enough to trigger another piece of the same iceberg to fall off after he plays the sound back to the iceberg. The installation in the gallery is supposed to demonstrate something like this. "It has to be *louder*," he complains about my last attempt.

"All right, then," I say. "Just turn it up. And get the biggest speakers you can find."

"It has to shake the whole room!" He's excited.

"Like I said, bigger speakers." Seems obvious to me.

"And make an even bigger piece of the ice fall off! Why not try dynamite?"

"Doesn't the sound of an avalanche trigger another avalanche?"

"I don't think so."

"But it *could*, no?"

"Possibly. What is this supposed to be, a kind of metaphor for humanity destroying nature's ice reserves? That's already happening, but not because of any big noise."

"Well, it's art. Drama. A photographic spectacle."

"I see. It *could* work, at least be stark and compelling. Better than some of what I saw at the Hamburger Bahnhof the other day. There's actually a piece there called *Shit Head*, which is nothing more than a sculpted likeness of the artist made out of his own feces and encased in several layers of Lucite so we don't experience the smell. Though maybe it would be more of a spectacle if we *could* smell it."

"At least there would be less doubt as to what this work is."

"Roger Shattuck once said that 99 percent of all art is shit."

"At least 100 percent of this one artwork *is* shit."

"But we're talking about it now. Doesn't that make it important?"

"That's what Arthur Danto said, may he rest in peace, but I still think there ought to be something more, something beautiful to take in."

"At least we agree there. And yet I wonder ... why was the most moving piece of contemporary art I saw in the entire Bahnhof simply a large dead owl lying in a box of salt? I can't stop thinking about it."

"Our sense of beauty works in mysterious ways."

"Like everything else, I guess."

"It's hard to top Joseph Beuys's oblique speakers blurting out 'Ja ja ja ja' and 'nein nein nein nein.'"

"That almost says it all. The nonsense root of human language."

"The absurdity of imagining any sound you hear can mean anything at all?"

"The blah blah blah of human existence."

"I also saw dead dogs and dead rabbits in that museum. And a room full of crude clay birds on a plywood floor. Don't step on them! Don't touch! Don't go inside the green wooden frame that makes you want to walk inside it! *Nein nein nein nein.*"

"*Ja ja ja ja.*"


I can no longer tell who is speaking each line. Art can do that. The memories of the conversation just swirl in my head. A bird in a cage is not worth more or less than one in the wild, but it is a different animal altogether. There is a magical film of the multi-reedist Rahsaan Roland Kirk (indeed, he would sometimes play three horns at the same time) wandering with a flute through the London Zoo, a toddler on his shoulders, dark shades on, testing melodies along with lions and geese and bears. You hear a jazz of give and take, outside and inside the cage of improvisation and tradition. The critters seem to like it, and who can blame them? With all these human gawkers empowered by the bars to leer at and taunt the animals, finally someone is trying to engage, and with the power of music no less, a message of emotion and beauty that goes where no words dare tread.

With this famous film of the great jazzman and the tigers, I accepted a proposal from the Grey Cube Gallery to visit the Helsinki Zoo, on the island of Korkeasaari, to engage with their animals inside and outside the aviary early one June morning, before most humans were allowed on the scene.

In the bird hall I found my old friend the white Bali starling, a beautiful species of mynah now thriving only in the world's aviaries and zoos. His song is hoarse but melodic, awfully hopeful for a bird nearly extinct in the wild. I take out a pure-tuned *seljefløyte*, a Norwegian overtone flute that plays only the natural harmonic series, with no finger holes to eke out any human-tempered scales. It just offers up the wavering pitches of the wind. The Balis seem to like it, along with the other colorful finches and thrushes that surround them. Just like them, waking up to the day inside the enclosures of man, bounded yet more or less free, I play short snippets of tentative song. I wonder what world they are missing, their wild compatriots all but gone from the planet.

I've been joined by my good friend Petri Kuljuntausta, a chronicler of the history of experimental music in Finland and a master of unusual electronic sounds. He plays brief invocations on an electric guitar, glancing up to the glassed-in trees.

Later I head outside and find myself face to face with wailing peacocks displaying magnificently all around me, lifting their beaks up into alien howls. I shriek along; they join in tandem. One spreads a powerful display. They march around the lawns. I remember the ornithologist Richard Prum telling me, "Most of those peacocks never even get one chance to mate," and I wonder if that's all this magnificent excess of evolution really is: sheer frustration.

I walk through the habitats, clarinet held high, playing something rather than nothing. The lions and crocodiles gaze out with indifference. I try to stare down a takin, a rare Himalayan moose not often seen in (or out of) captivity. John Berger asked, "Why look at animals?"—he was troubled by their mute, unfeeling gaze, which seemed to penetrate us so deeply at times.[3] Do they see through us? Are they so hard to impress? More so behind bars than in the wild. There is always a deep sadness in zoos, and yet both children and adults love them so. They will not go away; humans seem to want animals close enough for observation, with all this mystery stored away in safety. In simplest terms, the animals are like us and unlike us, with agency and curiosity, yet no existential remorse or lingering anxiety about to question everything.

In their music I have always found a kind of certainty that human creativity by its very nature must lack. We are never sure what song to sing or how exactly to sing it. We do not know what is good or bad, and these days are afraid to even ask that question. They sing or are silent and may not trust us, but they know exactly how to live their lives. And since they are alive, we know in some sense they are just like us, and we like them, part of the lasting world. Even if mute they know us in a way we can't know ourselves.

Wild whales, birds, and bugs clearly sing from their natural habitats, while the starling and peacock seem stuck in a human place; there can be a music with things that falsely seem to be alive. In zoos, writes Berger, the animals disappear—into their own irrelevance and detachment from our hollow lives. They used to live and work with us in our houses. Now we just go to watch. To look. They stare at us in mute incomprehension. Or uninterest, or contempt. Hopeless contempt.

Zoos—sad places of delight. Laboratories of experiment when we try to play along. We jail or cage the ones like us but not quite like us. Back in nightingale city there is a newish mall called Bikini Berlin built right next to the Tiergarten Zoo. A large plate-glass wall faces the gorilla enclosure. There the apes can watch ridiculous specimens of humanity poring through sale racks at pop-up stores, engaged in that pointless activity called "shopping." We laugh at them when they roll apples in the dust and pop them in their mouths. They may laugh at us in kind when we look at cryptic marks on price tags to decide if these are things we need, or even want.

I watch myself playing along with peacocks and starlings, at once happy and sad. It is sad to see animals trapped, but I laugh when they seem to join in. It is nowhere near as fun as that original laughing thrush that got me started on interspecies musicking many years ago, but the zoo experience presents the context, the severity and the relief. What is not recorded is the fact that in the morning we went in with the creatures before the public was allowed and engaged with them on somewhat their own terms. Later in the afternoon the public was invited in, and we were part of the show. I stood in the shadows with my *seljefløyte* and tried my best to disguise myself as a bird. To feel like a captive resident of this place, on display. The spectacle was announced, and all these people rushed in.

Phone cameras aimed at the ready, snapping, flashing, recording, social-media-marking a moment of hopeful interspecies contact. Set up the possibility for music, and the music may appear.

Everyone watching and wanting, ultimately by the boredom all zoos eventually make us feel. The animals are stuck there, but we leave, and wonder why we caught them and what they mean to us.

I want to play music with these musicians of other species, not to taunt them but to engage them. I don't want them to copy or challenge me. Naively, perhaps, I imagine we might make something beautiful together that neither of us could completely understand, in a sharawaji of many species coming together as one multifarious sound. With this vision of musical life, everyone contributes their own beauty, magic, emotional form, or mystery, and together they approach the inexplicable. I hope it may work, so I travel the world finding opportunities to do it. All the while I am thinking of the nightingales of Berlin, when I will return to them, when I can be out in their forests—not trapped behind a Bikini Mall wall, rifling through my own human obsessions. There is a possibility, albeit a faint one, of tuning into this ancient and lasting music that has been on this planet for millions of years. For that reason alone, something about it should matter.

# 3

## BEGINNINGS OF TIME

Soundscape pioneer Bernie Krause reminds us that hardly any-where on Earth is free of human sound. Our mechanized dins and hums are everywhere, and we cannot escape them no matter how hard we try. Yet in crisis, there is opportunity. At least that's how the nightingales of Berlin must see it, for one must wonder why this most international city, straddling the border between the old East and West, is so full of these enchanting singing birds. There have always been birds that thrive in human-altered landscapes such as fields, lawns, and parks—robins, starlings, house sparrows. The nightingale, though, has a far more remarkable song than any of these.

I improvise with these other species just as I do with human musicians with whom I might not share a spoken language. I have learned much from the ethereal, open-ended improvisation of many of the musicians recording on Germany's ECM label, includ-ing the saxophonist Jan Garbarek, the pianist Keith Jarrett, and especially Don Cherry with Colin Walcott and Nana Vasconcelos as the band Codona. I prefer those performances most freely impro-vised, yet still somehow rhythmic, simple, and listenable. Perhaps that makes me more of a populist or folk musician, but I still hold to the idea of "sudden music," the name of a book I wrote at the end of the twentieth century, trying to figure out exactly what it is that only improvisation can create.

In improvisation you can hear the greatest personality of the performer, more than his notes or his instrument. There is a depth in simplicity, a value and weight in paying attention. This is what my friend, the late composer-performer Pauline Oliveros, meant when she spoke of *deep* listening: "Listen to everything all the time and remind yourself when you're not listening. . . . We cannot turn off our ears—the ears are always taking in sound information—but we can turn off our listening."[1]

To deeply listen to a nightingale is to feel the power of a musician who is not human, a purveyor of sounds as ancient as they are futuristic. At once primeval and electronic, these whistles, scratches, and turnarounds are inarguably music. You can prove it or you can feel it, whichever approach works best for you. The nightingales converge on Berlin as much as the artists, the travelers, the seekers and slackers, entrepreneurs of a waning belief in multiculturalism and globalization, so that it is possible to come together and make something new and beautiful. That said, the possibility of hearing it is easy, for it has always been resounding all around us in literature, history, myth, and story.

The nightingales really are in Berlin, all over the city on those first warm days in late April or early May, having recently flown in from Africa via the Camargues, set to recover their old territories from the year before or to prospect for new ones and further develop their lifelong songs. People do hear them. We all hear them late at night, just before dawn. Even at any hour of the day one might sing out, celebrating the dawn, dusk, or night, proud to be alive and destined to sing.

This special bird's emphatic song caught the attention of poets centuries before musicians knew what to do with his bleeps, whoops, and rhythmic clicks. Samuel Taylor Coleridge wrote of a time in England when the nightingales were far more numerous than they are now, and he noticed one secret booming sound:

> So many Nightingales: and far and near
> In wood and thicket over the wide grove
> They answer and provoke each other's songs—
> With skirmish and capricious passagings,

And murmurs musical and swift jug jug
And one low piping sound more sweet than all ...

In his tenacity the nightingale stands out from all those other won-drous singing birds. He does not perform alone but waits for an-swers from his kind, and is not afraid to answer us. He may not care for us, but he will acknowledge our musical presence, with that one low piping sound sexier than every other. He knows he will win. He will always win. No matter how long we play, he will outperform us.

Coleridge's compatriot John Clare spent far more time laboring out in the fields and knew the nightingale's song as something tan-gible and real. He listened to it and felt its rhythmic power, which he posits as the possible origin of human poetry in his manifesto, "The Progress of Rhyme":

One moment just to drink the sound
Her music made, and then a round
Of stranger witching notes was heard
As if it was a stranger bird:
"Wew-wew wew-wew chur-chur chur-chur
Woo-it woo-it"—could this be her?
"Tee-rew tee-rew tee-rew tee-rew
Chew-rit chew-rit"—and ever new ...

Words were not left to hum the spell
Could they be birds that sung so well?
I thought, and maybe more than I,
That music's self had left the sky
To cheer me with its magic strain
And then I hummed the words again,
Till fancy pictured standing by
My heart's companion, poesy.

This is a poetry of sound, not words, of animals that include the human and offer up their storied possibilities to us. It isn't just the energy that grabs him, but the specific and actual sound that humans tend to forget as we name the facts of nature away. Let listening sometimes be forgetting the names of the birds we hear. This man knew his nightingales well enough to try speaking their language and thereby discovering the music hidden within it.

Percy Bysshe Shelley knew not everyone loved these birds, or nature at all. He was also aware that great Persian singers were often called "nightingales," and he wanted to be one, too: "A poet is a nightingale who sits in darkness and sings to cheer its own solitude with sweet sounds; his auditors are as men entranced by the melody of an unseen musician, who feel that they are moved and softened, yet know not whence or why."[2]

In "The Woodman and the Nightingale" he offers a parable of the battle of man versus nature, or at least poet versus pragmatist:

Like clouds above the flower from which they rose,
The singing of that happy nightingale
In this sweet forest, from the golden close

Of evening till the star of dawn may fail,
Was interfused upon the silentness;
The folded roses and the violets pale

Heard her within their slumbers, the abyss
Of heaven with all its planets; the dull ear
Of the night-cradled earth; the loneliness

Of the circumfluous waters,—every sphere
And every flower and beam and cloud and wave,
And every wind of the mute atmosphere,

And every beast stretched in its rugged cave,
And every bird lulled on its mossy bough,
And every silver moth fresh from the grave.

With a line like "interfused upon the silentness," Shelley has attempted to translate what each lilting phrase and breakbeat might mean if shifted into English. The midnight avian song contains all these possibilities, said and unsaid, wondered and unremembered. But the woodman himself *can't stand it*:

> Whilst that sweet bird, whose music was a storm

> Of sound, shook forth the dull oblivion
> Out of their dreams; harmony became love
> In every soul but one.

There's one naysayer in every crowd, is there not? He isn't going to let that hopeless unstoppable song prevent him from cutting down their homes so people can have wood to burn and build. The birds will need to sing somewhere else. The crass woodman is not sympathetic to the "mute persuasion of unkindled melodies." Let the ax come down:

> Wakening the leaves and waves, ere it has passed
> To such brief unison as on the brain
> One tone, which never can recur, has cast,
> One accent never to return again.

> ...

> The world is full of Woodmen who expel
> Love's gentle Dryads from the haunts of life,
> And vex the nightingales in every dell.

Such people, destroyers without conscience, are everywhere. Yet is that not what we musicians are also doing by vexing the nightingales in every dell? Hence the title of *And Vex the Nightingale*, a trio album released in 2015 featuring one virtuosic bird, the singer and composer Lucie Vítková, and me on clarinet and iPad.

It all began with a rehearsal one midnight in May of that year for a concert to be held the next night. We searched Treptower Park for that one special bird next to the frog pond that had sung so intensely the year before. We practiced for nearly an hour with this solitary bird, by the shores of the Spree, the S-Bahn rumbling in the background and occasional night revelers laughing in the park. I was so impressed with Lucie's sensitive, pure tones, gently leaving space for the phrases of the bird, and the looming simplicity of the hour, that we released it that winter.

What I like is that the bird holds center stage. No one tries too much to imitate or to flummox him, but instead we try to make music with him in the mix, even through sampling him live or transmuting his tones into electronics. To *vex* the nightingale: what a wonderful word that no one uses. I guess it means to goad, to nudge, to annoy—and maybe we are doing that to our *Luscinia*. But maybe he is just doing his thing faced with a hitherto unheard music in his realm. Then again, this *is* a park in Berlin; he's probably heard more than his share of surprising sounds. And he must like it, because every year he comes back to this same spot.

Nightingale music takes time. He can go on all night; we should be able to manage one whole hour. This performance *must* be one long unearthly, untaught strain. Listening back to it all now, I can tell it takes time to get into this nightingale scale of midnight hours—sound weaves slowly around my consciousness, taking forms and shapes only after many repeats start to make sense. Bird, voice, and clarinet: each marching its way across the minutes. Composer Olivier Messiaen said birds are the "opposite of time," because they have always been here and always will be, but in this immediate night, the 'gale marks time and its openings, leaving space for whatever ambience rests, or whichever musicians pass under his darkened thicket perch nestled strategically in the tree. That night no frogs decided to join in.

It still seems incredible to me that these birds come back, thousands of miles, to the very same tree they defended as territory the year before. Scientists and musicians who pay attention to these things can tell one bird from another by the number of songs they know and by the quality of those songs. This particular bird, identified innocuously as "#7" on Silke Kipper's Treptower Park nightingale territorial map, stood out two years in a row as having a particularly intense song. This is the bird she believed we had ruined for her, and for science as a whole.

Lovers have for centuries known well that the nightingale is the bird of love and lingering, lasting full on through the night until all the other birds chime in at dawn. Even Shakespeare's Juliet wonders if it's the nightingale who still sings beyond her dreams. Romeo corrects her—it is already dawn, and too late for their forbidden love:

*Juliet:*
Wilt thou be gone? It is not yet near day.
It was the nightingale, and not the lark,
That pierced the fearful hollow of thine ear.
Nightly she sings on yon pomegranate tree.
Believe me, love, it was the nightingale.

*Romeo:*
It was the lark, the herald of the morn,
No nightingale. Look, love, what envious streaks
Do lace the severing clouds in yonder east.
Night's candles are burnt out, and jocund day
Stands tiptoe on the misty mountain tops.
I must be gone and live, or stay and die.

Believe it: he is the one who keeps the music going as we gaze into each other's eyes, whoever we are in the parklands of the night, where nature meets the noise that never quite goes away.

It is important that he starts and stops. It is crucial that he signals the heart of spring, the beginning of summer, and that he goes quiet as the warm days go on. Shakespeare continues his reverie in Sonnet 102:

My love is strengthen'd, though more weak in seeming;
I love not less, though less the show appear:
That love is merchandiz'd whose rich esteeming
The owner's tongue doth publish every where.
Our love was new, and then but in the spring,
When I was wont to greet it with my lays;

As Philomel in summer's front doth sing,
And stops her pipe in growth of riper days:
Not that the summer is less pleasant now
Than when her mournful hymns did hush the night,
But that wild music burthens every bough,
And sweets grown common lose their dear delight.
Therefore, like her, I sometime hold my tongue,
Because I would not dull you with my song.

Let's forgive Shakespeare for not knowing that it is only the male nightingales who sing. Nearly every Romantic poet still made the same mistake a few centuries later, when they should have known better!

Of course, the nightingale is supposed to be Philomela, a princess from Greek mythology. Raped by her sister's husband, Tereus, the king of Thrace, Philomela threatens to tell her sister, Queen Procne, of the abuse, whereupon Tereus cuts out her tongue. Unable to speak, Philomela weaves a tapestry that tells the tale. When Procne receives the weaving, she plots revenge on her rapist husband by killing their son and tossing the boy's head onto the dinner table in front of him. Before he can continue his violence against the two women, the gods turn everyone into birds: Tereus into a hoopoe, Procne a swallow, and Philomela a nightingale.

So the nightingale has been heard by some to sing of sorrow and violence for a thousand years, the unspeakable horrors of which humans are capable. He bursts and halts, giving us time to think, to reinterpret, to insert our own visions of this inscrutable song. He is a bird so obsessed with love that he pricks himself and bleeds to death upon the thorn of the reddest rose. He comes to us with messages of desire. I'm with Angelo Badalamenti and David Lynch, who have singer Julee Cruise sing in *Twin Peaks* of a heart flying through the night with a nightingale all across the world, looking for love. Like a lover's heart, the nightingale returns to the same branch, to songs always the same and never the same, ever developing, ever bending, ever remaining beyond the reach of our translation and interpretation. Only music crosses the line from the human species to another. With sound and sentiment, I'll keep trying.

There is no need for music to last in this age of endless virtual sound. Those who make it need to keep showing up, doing the same things over and over again to keep them alive, to make them matter. It can be tough to continue going through with it.

I remember taking a trip to Hawaii with the great writer Rick Bass for him to chronicle my efforts playing music live with whales. Every day we went out, over and over again, hoping to find some singing whales and a moment when my clarinet would interact with them. Their songs are like long, drawn-out, slowed-down nightingale tunes, equally extreme, like-minded outliers on evolution's tree. Why such ornate singing by animals separated by millions of years of selection? We don't know. But extremes sometimes make the best examples, or at least the best music. It is our duty to engage with them.

So we are taking a break, at the very southwest tip of Maui, walking the famous basalt path called the King's Road, upon which few people venture because it wanders off to the rough shore, where it ends. No beach, no softness, just rough jagged rock at the edge of the sea. Not so popular. Our kind of place.

Bass is a mountain guy from Montana, where he lives in a distant valley called the Yaak, far from everything except the purest natural truth. Here in Hawaii we are awash in a different kind of nature, a place initially foreign to almost everyone but soon easy to love. We have met so many people who came to these islands and reinvented themselves, choosing a new name, a new job, a new vision of themselves and what they could do. He has almost had enough of this tropical fantasy.

"I don't know, David. Can we please *not* meet yet another beautiful middle-aged survivor who has changed her name to Moonglow, Starwater, or Bliss? I just can't take another story like that today."

I laugh. We continue in silence as the hard rock cuts into our sandals. We glance out at the rough sea to discern the spouts of whales from simple whitecaps. Not always easy to tell.

"I don't know if I can keep going in this writer's life," he muses. "I just turned fifty. It's hard to get so many words down anymore. And it's hard to find more readers beyond my dedicated fans. I should have taken one of those teaching jobs they always used to offer me when I was at my peak."

"Relax, Rick. Your peak lies far ahead, way beyond the other side of these waves." The crash and the sun and the wind seem almost perfect. Never mind the whales. We are entering sharawaji territory right here on land.

"You know, we should go visit Merwin. W. S. Merwin, our former American poet laureate. I know he lives on this island somewhere."

"You know him?"

"Not really, but I did call him and warn him we might show up. He has read some of my writings on birdsongs."

"Really?"

"He said sure, stop by. But call first. And I quote: 'There is no way you will find my house through the thick palm forest that surrounds it.'"


Bill Merwin met us at the top of the lane at the arranged time. He was right: there was no way to even imagine there was such a house down in there. In his mid-eighties at the time, he was excited to talk of nightingales, and of possibility. We discussed poetry and the special trees he grows on his property. How great it was to learn that he had written a poem that echoed what he read in my earlier book *Why Birds Sing*. What an honor to hear him recite this part of it:

> Long after Ovid's story of Philomela
>     has gone out of fashion and after the testimonials
> of Hafiz and Keats have been smothered in comment
>     and droned dead in schools and after Eliot has gone home ...

after the name has become slightly embarrassing
    and dried skins have yielded their details and tapes have been
slowed and analyzed and there is nothing at all
    for me to say, one nightingale is singing . . .


after the recognitions and the touching and tears
    those voices go on rising if I knew I would hear
in the last dark that singing I know how I would listen.

Deep among his beloved palms, hidden in plain sight on a Hawaiian isle, Merwin listened with optimism, belief in the power of words and of art, and faith that humanity won't lose touch with all that wonder around us.

He was way past his own midlife crisis, and even the midlife crisis of our species. The nightingale, he knew after a long life of thinking about such a song from his years living in rural France, will always be more than our interpretations of it, and will endure past the demise of our species, along with all we claim to control and know. Once our merely human songs and poems and myths of the nightingales are long gone, the original song itself remains as necessary and enigmatic as ever.

This song—I want so much for all of you to hear it, not just recorded on the etherwaves but in person, in the flesh, in dark and in light, fleshed out through a history of innovation, variation, and eternal tradition. A music like no other, but one we can reach if we are ready to listen. Merwin listened for years. He knew, he remembered. He read the science and smiled at the science. He assessed all that poetry and knew its brilliance and its limits.

Rick Bass, I am now older than you were back then when you questioned where all this was going. I get it, and feel where you were. And Merwin? He's still out there, dreaming of all those projects he may never get to finish with the same need to get there. May we all have such dreams and visions if we reach so venerable a place.

# 4

## ORDERLY AND DISORDERLY

The science and art of nature do not always mix easily. I tend to speak of these two "cultures" in terms of what they value: science and art have different criteria for truth.

I can play a duet on my clarinet with a nightingale in Berlin, connecting magically in the night. The resulting encounter may sound beautiful on a recording, and I can offer it up to the world for listening and enjoyment. I am elated to have been a part of this single field-recorded event, and to share it with all of you if done well.

A good scientist would ask: how do you know you are *really* affecting the nightingale? You would need to do this hundreds of times, playing control sounds to the bird to establish a baseline for consistency and avoid familiarity bias. Otherwise, it can't be a viable scientific experiment. There must be enough data, rigorously collected, and solid statistical analysis to back up any claim for interaction you want to make. That is the scientific criterion for truth—statistical probability that your hunch is correct.

The artistic criterion for truth is that of John Keats: beauty is truth, and vice versa—just give me something thrilling and I will smile. All it takes is one touching example of people and birds making something enthralling that neither could create alone.

Philosopher Karl Popper believed that science only traffics in falsification, not verification, although no scientist likes to hear this. Does this make aesthetic success so much easier and more certain than scientific validity, where we are never sure but must make do with probability?

Whether in regard to a dance based on the discovery of gravitational waves or a symphony written upon the rules of fractal mathematics, so often I hear from frustrated artists disappointed with scientists who refuse to value their revolutionary ideas. Conversely, there are scientists who offer up beautiful images to exhibit on the walls of art galleries, only to find themselves equally rebuffed. Each side must learn to play the other's game if it wants to be welcomed.

While there are plenty who do make the leap, it is difficult from the outset. I don't know if there really are two cultures, but there are two ways to become sure of your ideas. Don't read science as you would read art; the words are rarely mellifluous or inspiring, but the ideas can be magnificent, and should be presented efficiently. Don't look to art to explain what's right or wrong with the world. Listen, touch, and see. Metaphors from science can easily inspire art, but they don't necessarily lead to insights that scientists can use. They don't have to.

When I started this kind of writing fifteen years ago, I spent too many words criticizing scientists. It bothered me that they were afraid to assess the actual beauty of birdsongs. Later I was impressed that in their pioneering report on the structure of humpback whale songs, Roger Payne and Scott McVay wrote that the humpback whale "emits a series of surprisingly beautiful sounds."[1] No subsequent scientific paper on animal vocalization has since dared to call such sounds beautiful, because beauty is not a scientific parameter. Almost everyone, including scientists, finds birdsongs beautiful, but specialists struggle to measure their beauty.

In his poem *Lamia*, John Keats chastised scientists for being the kind of people who would "clip an Angel's wings, Conquer all mysteries by rule and line." Biologists who study flies actually do sometimes pull off their wings and put the little insects on tiny wheels to measure how they try to move.[2] But rare is the animal-rights activist who'll get worked up over bugs. Such examples, however, are the exception to the rule; most scientists are kind to the animals they study. They love the songs of birds and whales and insects as much as the rest of us.

Sometimes I am suspicious of that part of human intelligence capable of weighing us down in the face of tremendous beauty or untouchable nature. What can we really say of this essential world from which we come? We all know it is more beautiful and necessary than anything our species can add to it. Yet we constantly change it, ask it to divulge its secrets. No matter how much data we find, how much we read into or examine the world, the value of our initial meeting with the perfect and the wild will not be diminished. And while information can (and does) diminish the apprehending of beauty, we must try not to let that happen. Hold all we know at bay so that we may love the world.

No one way to know the world does it all. We need science as much as we need art, and each of us can use both to gird experience. Here I lead up to the ways my friends and I confront the nightingales, gently asking them through the emotions of sound how we all may make music together, thereby adjusting our honest place in the world. In the most important part of this work, the making of music with the resounding natural world, I do not expect any answers beyond the celebration of special and surprising encounters. Still, the science of nightingale song is an important part of preparing for these encounters, so be prepared for diagrams, charts, statistics, and uncertainty.

Silke Kipper has been leading the research group on nightingale song at the Free University in Berlin for several years since the retirement of its founders, Dietmar Todt and Henrike Hultsch. Kipper and her team have tried Markov chain analysis, network theory, and all manner of statistical analyses to compare one bird with another. Each nightingale in his prime knows from one hundred to three hundred distinct phrases or songs. Some birds sing in an "orderly" fashion, wherein certain sequences of phrases seem to regularly recur, while other birds sing in a "disorderly" or chaotic fashion, in which no sense of order can be discerned. What is the difference? Does the more organized bird know more than the less organized one? Or does he only have a different style? So far we don't really know.

Scientists like Kipper try to decipher the birds' scratching and sampling with as little reference to human musical categories as possible. That is because she and most scientists call our musical ideas humanly subjective, and thus less relevant to science than the penchant to count, measure, and enumerate. She and most scientists feel that enumeration is as close to the sound in and of itself as we can get.

Music, science, poetry—three separate yet equal human ways of knowing. None of them on its own will completely explain what it is like to be a nightingale. When Thomas Nagel challenged us with his paper "What Is It Like to Be a Bat?" he concluded that we will never be able to know the world the way bats do.[3] Logical, perhaps, but no argument will stop scientists and musicians alike from trying to see, hear, and play the world in the manner of other species. We remain fascinated by all these alternate animal ways of singing meaning out of this world because these are the expressions of beings beyond our own species, who seem to enjoy some of the same things we do.

Scientists are by their creed far more impressed with order than disorder, and thus Kipper is worried by the occasional penchant of the male nightingale to insert a strange "buzz sound" into his clarion whistles, clicks, and ratchets. She admits she finds this buzz unpleasant—not that her opinion matters, as the female nightingales find it especially pleasant. I had noticed this sound as well, and that the nightingales would sing it only occasionally, like an ornament, a grace note, or even a blue note—that cool, unclassifiable sound human music is known to offer up in many forms, a hip tone, a most excellent riff that only makes sense if used sparingly.

Hear that buzz amid all those whistles and clicks! Scientists like to print out sounds we cannot easily describe. These images, like the many others to follow in this book, are called "sonograms" because they visualize sound, mapping frequency against time. The more you study them, the more sense they begin to make. For now, think of that one fuzzy sound as a buzz, whirr, scratch, hook, or riff that stands out from the other sounds. *Brrrrrrrrjjrrrrrhh!* From now on I'm going to call it the *boori* sound because it's neither buzzy nor bluesy, neither ugly nor human. Decide for yourself whether it is music or noise.

Would the nightingales use it all the time if they could? Kipper

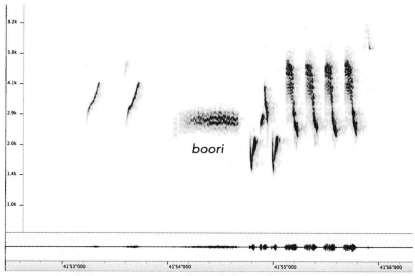

FIG. 2. The sexiest of all nightingale tones, the *boori* sound

suggests that this sound must be stressful for them to make and therefore requires virtuosic control of the syrinx (which, unlike the larynx, allows birds to produce two tones at once). When a female hears this, she knows that this male singer is strong, solid, and a good choice for mating. It gets her excited. She just might fly right into the nettles to find him ...

Through various playback experiments, Kipper and her collaborators established that the female nightingales like this *buzz* sound far more than any other note the males can belt out.[4] "Then why," Kipper wonders, "don't the males just sing this 'sexy' sound over and over again? Do they not know how well it works?"

I had to laugh a bit when I first heard the sound. It's like a bluesy riff, an out-of-tune quip, a wah-wah pedal tone. These effects are cool, to be sure, but any guitarist knows one shouldn't play them all the time. You've got to hold in your best licks, only letting them out when your audience least expects it. Keep that idea in your head when you listen to nightingale song, and an entirely different vibe emerges. It's something you can feel, and as a scientist you can also measure it.

I've heard all sorts of explanations for the lure of the *boori*, the uncanny, the in-between, the wrong that is exactly right. There are so many ways to ground sound in the ineffable—the breathy pillow underneath a jazz clarinet sound, the gliss, the bend, the characteristic slide in Gershwin's *Rhapsody in Blue*. You simply cannot do this stuff all the time. Music is an exercise in contrast between the expected and the unexpected, the beat and the stop, the patterned and the patternless.

At first, scientists laughed at me when I suggested they look into nightingale aesthetics. Their tendency was to dismiss my concern as the naive view of an outsider philosopher and musician. Aesthetics, they surmised, is an exclusively human affair. I reminded them that it was Darwin who emphasized that birds, with their females' ornately coevolved preferences for outlandish traits and behaviors, can clearly be said to have a "natural aesthetic sense." Although such ideas have largely been ignored by biology until the recent work of Richard Prum,[5] the Finnish biologist Olavi Sotavalta, who also had a degree in music from the Sibelius Academy in Helsinki, was the first to try this out fifty years ago on the thrush nightingales of his native Finland (see plate 1).

Sotavalta wondered if he might apply musical analysis to a genuinely complex birdsong, that of the thrush nightingale, *Luscinia luscinia*, a bird of Eastern Europe and Asia with a scratchier, more rhythmic song than the *Luscinia megarhynchos* nightingale of Western Europe's romantic poetry. English nightingales sing fifty to two hundred different phrases with much variation and change. Sotavalta noted that thrush nightingales sing an equally large number of distinct phrase types, but that each type has a fairly consistent structure, more stylized than the phrases of the more famous bird, so he could categorize their variations much more succinctly. He listened intently to his nightingale's song. Its timbre is not harmonious, but raw and complex, combining percussive rhythms and clear notes: "Pure tones could be whistling, piccolo-like, dull, like a low flute, metallic, celesta-like or chippy, like a xylophone, long or short." He struggled to put it in words.

FIG. 3. Olavi Sotavalta's structural analysis of the song of the thrush nightingale

"The commonest noise-type appeared in the cadence and re-
sembled the rattle of a tambourine."[6]

Sotavalta studied two birds, one in 1947 and the other in 1948.
The first had fifteen basic phrases, the second seventeen. At the
level of the phrase, a definite form can be identified. In the thrush
nightingale song, the rhythm seems more significant than the
pitch. Figure 3 shows the basic structure Sotavalta identified that
fit nearly every phrase of both birds.

What is most significant about this mid-twentieth-century scientist's analysis is that he goes immediately, without apology, to
musical terminology. The introductory sounds are one or two soft
whistling tones, followed by a low-pitched antecedent, then a brief
link to the characteristic motif, which is the part most distinct between one phrase and the next. There are phrases in double and
triple time, at sometimes distinctly wide intervals, then a post-
cedent series of repeated low notes, a high *bleep*, final "chippy,"
xylophone-like chords, and that one quick tambourinelike rattle:
*chhuum*, our *boori* sound by another name. (Little did he know
this would be revealed as the sexiest of nightingale sounds many
years later!) Thrush nightingale deciphered? At least some structure is found. The revelation of a drumbeat music more resembles
a battery of percussion than a luminous turn of melody. With all
the praise given to the nightingale's virtuosity, it is amazing how
alien his music looks and sounds.

Sotavalta listened acutely and perceptively to decode the structure of the thrush nightingale's song. He found clear rules in it,
yet no line of research was ever drawn from his conclusions. Later
nightingale researchers scoffed at his sample size of a mere two
birds—that, and his use of an unquantifiable musical argumenta-

tion. Yet he traced the secrets of nightingale music more accurately than anyone since.

Inspired by Sotavalta, the musicologist Dario Martinelli, the composer Petri Kuljuntausta, and I organized a Nightingala Festival at Sibelius's home in Kallio-Kuninkala in 2008. This remarkable gathering brought together biologists, composers, musicologists, and performers to celebrate the musicality of nightingales and discuss ways our disciplines might collaborate. There I managed to convince the neuroscientist Ofer Tchernichovski that nightingale aesthetics could, in principle, become a subject that might be quantified in a scientific way.

Tchernichovski is famous for crunching the greatest amount of data of any birdsong neuroscientist, long before "big-data" analysis came into vogue. While some of his colleagues are more prone to murder baby birds at the moment of their singing to examine exactly which neurons have fired in their brains, Tchernichovski instead believes in computerized *listening*; he has recorded every single sound zebra finches make in their first three months of life, the only time they are able to learn to sing, in what is called the "sensitive period." He uses an algorithm of his own design to identify recurring patterns in the song in terms of amplitude, time, pitch, and something called "Wiener entropy," which is a measure of the relative amount of noise in a single birdsong syllable.

With this kind of statistical quantification he has been able to discover much about how birds learn to sing: at what time of day they learn new phrases, how a full song coalesces out of individual parts, and how they need to forget some parts of the song in order to remember others. Remarkable research, with a lot of data to back it up, all the while harming as few birds as possible.

Tchernichovski works mainly on the zebra finch, the model species in birdsong neuroscience about which the most is known. Neuroscience is interested in birds because they are one of the very few species—including whales, dolphins, and humans—able to learn sounds beyond instinct. Not even other intelligent primates can do this. Why not? Natural selection must somehow select *against* vocal learning. It must not be advantageous to most species, because it so rarely appears. But it offers untold possibilities for communication and enhancement, and only in the rare cases of songbirds, cetaceans, and humans has it evolved.

Even though the zebra finch has a very brief and stereotyped song, the process of learning is complicated. By developing a way to analyze hours and days of singing, Tchernichovski has made real progress toward our understanding of how birds learn to sing.[7]

I asked him to consider applying this same statistical and quantitative approach to the analysis of a single nightingale singing one very long and complicated song for hours and hours through the night. He immediately agreed that it could be done and soon afterward found a very capable postdoctoral student straight from Berlin: Tina Roeske.

Biologists convert sound into image so it can be analyzed with greater ease by our very visual minds. Roeske decided to focus on the amplitude of distinct motifs in the nightingale's song, and she color-coded the data, red being loudest, blue being silence, along with a continuous visual spectrum of colors in between. She then plotted the continuous amplitude of each three-to-eight-second song phrase (she and most biologists tend to think of each of these phrases as a complete song, while as a musician I tend to consider the whole performance or "song bout" to be a single piece of music, which makes it much harder to analyze). She aligned each song phrase from the beginning to discover what common features each phrase might reveal statistically (see plate 2). The colors only illustrate the amplitude or volume of the phrases. The most visible pattern appeared at the very beginning of each song phrase. Here there are *hundreds* of individual songs being compared.

Next is a single picture showing 420 different phrases sung by a single bird, k31. The color makes it possible to look at so many songs all at once. The x-axis is time in milliseconds, making each phrase between three and eight seconds long (see plate 3). The whole performance could go on for many hours. What does the picture show? The beginnings of the phrases have a common rhythmic form, followed by a dot-dash kind of order that resembles Morse code. Otherwise, a kind of secret confusion follows forth.

Science needs to represent animal music as statistics and graphs for these sounds to be taken seriously as data. As a musician I am happy to enjoy their songs as music, slowing them down to better appreciate their nuance, playing along in the studio or out in the wild with the real thing, to get inside the bird from an aesthetic perspective.[8] Science and music are two distinct forms of human knowledge, and they take disparate approaches when confronting the enigma of complex birdsong. Musical methods can help reveal the beauty inherent in the form, inflection, and energy of these sexually selected songs. The insight of the musician can reveal aesthetic criteria to explain how certain songs contain the same qualities we identify as beautiful in human music. Music can suggest such formal and emotional elements, but it cannot prove that they are really there. Science can analyze empirical data to support or refute aesthetic hypotheses, but only when it admits that the components of the beautiful can be measured as something more than subjective opinion.

The sifting of so much song data into a single image enhances our ability to think like a nightingale. So much about each bird's repertoire of songs is distinct from that of all other individual birds that it is important to find one or two statistically similar features. These become the quantifiable features that might begin to identify, for humans, the particular species aesthetic for the nightingale, the basic rules behind the music female nightingales would consider acceptable or preferable.

What is the *purpose* of this preference? Roeske and her collaborators are hoping to find some qualities in the song that will correlate with more mating success, thereby discovering the details of some sexually preferred trait among the aesthetic experts of the nightingale species—namely, the female birds, who are doing all the choosing.

As a musician I am more interested in what rules might lie behind the music, so that I might learn from its form and structure. Roeske also demonstrated, albeit with a small sample, that certain rhythmic principles do indeed distinguish one bird from another. Figure 4 is from a presentation given at a biosemiotics conference in 2011 in New York. Each of the four birds in question has a specific way of handling the space between each syllable or motif.

A "stereotyped bird" is similar to the "orderly bird" Kipper de-

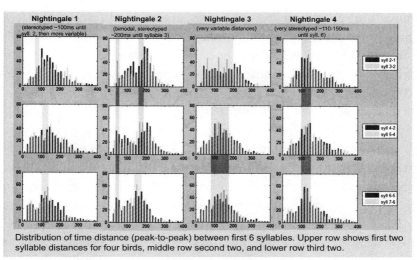

Distribution of time distance (peak-to-peak) between first 6 syllables. Upper row shows first two syllable distances for four birds, middle row second two, and lower row third two.

FIG. 4. Stereotyped versus variable performances by four different nightingales

scribes, while a "variable bird" is a "disorderly bird." Do this same sort of approach with human musicians and you will probably be able to find statistical tools able to distinguish one person's playing style, or one composer's writing style, from others' (how exactly is Coltrane's virtuosity different from Parker's?). This suggests another possible reason for the vast diversity in nightingale performance: an individual male might be identifiable by his personal singing style. These birds as musicians are distinct individuals, with different styles. That in itself is a major discovery.

Do we know if nightingales think like this? Not yet.

How much of a nightingale's phrases invokes repetition, and how much is innovation? What's the ratio between the two in *all* music, human or animal? Music seems to accept far more repetition than language, but why? Elizabeth Hellmuth Margulis has written a fine book on this very topic called *On Repeat: How Music Plays the Mind*. She is entranced by the question of why it is we crave repetition when it comes to music, either in the catchiness of a beat or our joy at hearing the same songs or pieces over and over again. It is clearly a neurological mystery. Our brains and bodies must gain something from rehearing the same assemblies of sound. It must bolster us, not bore us. Whenever I worry that my own music all sounds the same, I am heartened to learn from Margulis that at least one-third of Beethoven's oeuvre comprises reworkings of his own earlier ideas.

One of Margulis's most amusing experiments is to add random repeats to very abstract contemporary classical music, a genre that often tries to stretch our ears toward structures that offer little for us to hold onto and recall. Her data shows that nearly all listeners, including those who were composers and new music performers, *preferred* versions of the pieces in which random measures were repeated. We need to repeat, or we remember nothing.[9]

Always step into the same sound twice, and pause a bit before you begin. Leave space for the listener to reflect. Possibly the best advice to give to any performer or composer, and especially the interspecies musician. Learn from the nightingale. State your case and then step back to listen to whoever else is out there.

A nightingale's song is full of repetition and variation. A code we can easily identify but not easily make sense of. Phrases come that are similar but not identical. Each assembly of motifs in itself blends likeness and difference. The thrill of melody making form, an alien music but not a language because there is no need to decipher it. It is a music we can join in.

We can easily identify that the song is that of a nightingale. With a little practice you can tell a thrush nightingale from a common nightingale, a *luscinia* from a *megarhynchos*, a *Sprosser* from a *Nachtigall*. The first is rhythmic and harsh, the latter relatively ebullient and virtuosic. Still, it's quite difficult to tell one individual from another. Yet that is what the females must do.

As someone who likes to operate in the moment, I don't always need to know. It's phrase versus phrase, man to bird. I respond immediately to the melody I hear, its twists and shouts, steps and leaps, pauses and repeats. After reading science I am impressed by the *boori* sound, and by a long chain of *bip bip bip bip bip bip bip bip bip bip bip bip*. It presents its own challenge, like that lone bird Sam Lee and I played with in Lewes, whom we had to walk forty minutes to find through a mucky field. Even Matthew Barley carried his priceless cello along to join the show.[10] This bird was the last male singing in his copse, a fitful symbol of English melancholy.

Many scientists fear mention of the word *beauty*, but Tina Roeske is one birdsong scientist obsessed by it. She is still putting thousands of nightingale songs through the wringer of statistical analysis, identifying shapes, forms, syllables, beginnings, middles, ends, volumes, spaces, silences, consistency, and entropy, and displaying them all in beautiful color pictures of quantified analysis. It may look nothing like the way musicians might understand these phrases, but it is stunning nonetheless—the sheer plethora of data in visual form.

What was the one thing Tina did find out? Short answer: humans prefer stepwise motion in melodic songs. Nightingales instead like to break direction. Once we start going up or down we want to continue going up and down. But the bird goes one step up, and the next goes down. Jump up a sixth, and the next jumps back down an octave. Melodically, these birds want to confound human expectation. In some ways they are the Thelonious Monks of the avian song world.

Rhythmically, the beginnings of most nightingale phrases are consistent: a few long whistles followed by a contrasting series of trills and clicks. But after the preamble things get interesting as each bird seems to diverge. Some go on to whistles, others to contrasting beats. The thrush nightingale doesn't really have the sexy *boori* sound.

Although after the first few notes the way each song proceeds differs wildly from one to the next, in a single bird there is consistency. It is therefore possible to identify which bird is singing which song from the later syllables. There is no consistency among an entire species, but there is within individuals. The songs do distinguish one bird from another.

I find this conclusion significant. Tina does not, largely because of the same problem that happens whenever scientists think of studying birds with complex songs. The great song biologist Don

Kroodsma once warned a student never to do research on mocking-birds, because they will "ruin your data." We seem to have figured out exactly what every sound of the chickadee means, but the mockingbird? Sheer confusion. We can't even make up our minds whether they imitate car alarms or car alarms were programmed to mimic them. So what if we come up with tools to distinguish one nightingale from another? We still don't know what makes one bird's song "better" than another's, because no one has ever found any correlation in the nightingale between certain song qualities and mating success. Such is the holy grail of biologists' appraisal of birdsong: what can these boys do to improv their sexual game out there? It may be a silly, base way to judge the beauty of life in nature, but sometimes that's all we have.

I find in Roeske's research all kinds of evidence to bolster my gut sense that the best moments in music can't be explained. You have to buck the trend. Roeske has done that by becoming transfixed by the beauty of the nightingale, who surprises us by keeping us listening and wondering. His whole species might do it by wanting to go down the moment the melody goes up, singing things that surprise us, lead us on, make us believe it's still worth trying to play along with these birds, making music no one species expects.


I think of the first time I was out in Hasenheide Park at 1:00 A.M. in April of 2014. A singing bird hides in a thicket just a yard or so in front of my face. I do not play Brahms, like Beatrice Harrison, but instead try to play half-human, half-nightingale music that maybe no one will like. The more I do it, the more it transfixes me. Each season I come back to the same parks and the same trees on the same days with the same birds, hoping the nights will be warm enough for endless music to come.

I don't envy Roeske the task of collating her data for months, programming, thinking, developing innovative visualizations that claim to present the songs without prejudice or human touch. These color layers of lines do not take away the magic for me but show the challenge of transforming beauty into data. The wonder is that color helps us encapsulate data. It makes science all the more beautiful, and if you want to measure beauty you mustn't be afraid of hearing it.

Poets came first to the celebration of beauty in the songs of birds in the eighteenth century, gathering all sorts of rhythmic and linguistic ideas from the warps and woofs of avian phrases. Musicians came much later, only once the odd rhythms and tones and scratches of the nightingales' songs became acceptable as musical material. Scientists in turn took all these virtuosic utterances seriously only after sound could automatically be turned into image, allowing us to sit and stare at our leisure to decode organization and structure in supposedly objective visualizations.

What's changed in the last decade is the ease with which those images can be shaped and finessed. I've always found sonograms spewed out by my computers to be beguiling and beautiful, and now we can sift through thousands of them in accordance with different indices to catalog ever more layers of digital data.

When I first implored scientists to ask harder questions about the music of birds, to not just assume that the male with the loudest or longest song is going to get the most females, but instead to try figuring out what the best nightingale song would be by delving into the deeper aesthetic structures of what birds like and dislike, most researchers thought my questions were beyond the realm of quantitative analysis.

After five years of work analyzing at least six thousand individual nightingale phrases sung by dozens of birds recorded all over Europe, Ofer and I, along with Tina and several other nightingale scientist luminaries, finally published a paper suggesting that musical thinking could help birdsong neuroscientists. The result, "Investigation of Musicality in Birdsong," was published in the journal *Hearing Research* in 2013.[11]

By focusing on something simpler, not even the song elements but the *space between* each sound and the next, Ofer had a hunch that rhythm and silence might mean as much as the diversity of sounds. The colored lines below (fig. 5) are fine enough to showcase the great confusion and slight similarity in all those hundreds of song phrases, but how to compare all the rhythmic variations in one single image? Tchernichovski's preference for square grid dia-

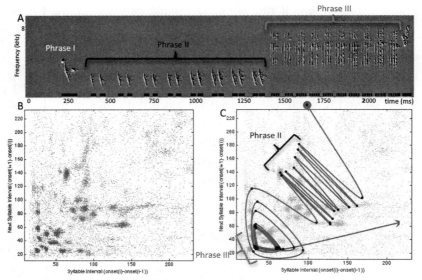

FIG. 5. The spaces between each nightingale syllable analyzed in a single picture

grams that map interval against onset may be harder for musicians to imagine as a score, but they are far easier to work with when amassing such large amounts of data in a single picture.

The red line maps one song as it travels through the phrase plot, while the gatherings of darker blobs indicate parallels in rhythmic silences between the sounds. It is clearly not random, but rather uses many possible rhythms. Keep in mind this species of nightingale, the *luscinia* of Helsinki, is much easier to analyze in this way than the *megarhynchos* of Berlin, with its hundreds of different songs.

And what of the different kinds of songs? There are whistles, clicks, and brrrphs. Here we map pitch (frequency) against "noisiness" (Wiener entropy, Ofer's favorite). This is a fragment of one bird's song, categorized into types and rhythmically scanned. Put all the phrases together in one grid and you get an analysis like the one shown in plate 4. Once again we can distinguish the organized bird from the disorganized one. Or is it that one can produce only a few specific sounds while another sings just about any which way but loose?

We still don't know whether orderly or disorderly is "better" from the female nightingale's perspective. Is the only way to measure this by determining who has more mating success? Maybe the best singers just sing and sing, making more and more elaborate music as the season blooms and ripens, precisely *because* they haven't been able to find a mate. At least that's the age-old story of a nightingale's singing for unrequited love, as so many human musicians are ever wont to do. Why should we expect them to be all that different from us?

And so, can bird music be quantified? As much as any other music I suppose, and with equal deficiency. "We could never find a pattern!" says Roeske, now back in Germany at the intriguing-sounding Max Planck Institute for Empirical Aesthetics. "I wanted so much for there to be a pattern, I got entranced with the beauty of the bird's song and forgot the impossibility of ever decoding it."

But in 2017 she did discover one strange secret of the nightingale's song. Roeske and her collaborators published a paper that argues that these birds swing. She tries to quantify exactly what it is about nightingale song that makes it sound so musical, and it turns out that the most quantifiable aspect of *Nachtigall* musicality is rhythm, not structure or form.[12] It don't mean a thing if you ain't got it, but swing is notoriously hard to define. It's that uneven syncopation that makes us want to move, not quite triplet and not quite three out of four. *DAH duh DAH duh DAH duh DAH duh*: if you can't feel it, you won't quite get it. We all know it when we hear it, that uneven evenness that makes you smile.

It turns out nightingales have this same quality in their rhythms, an evenness that is always a little uneven, in anything-but-random ways. Roeske has now tried something called "multifractal analysis," a technique previously applied to human music to measure specific kinds of unevenness that randomness will never explain. You don't need such numbers to get the bird's fine music, but finding such a musical quality bolsters our confidence that the jazz we hear in musical birds is far more than wishful human thinking. Science must count the uncountable, and that's how you measure the thing that means when you got that swing.

Why would a scientist need to apologize for being entranced by a bird's song? Science is by necessity wary of beauty, because it knows not what to do with that which it cannot contain. Beauty does not come easily to numbers or calculations. Beauty, like truth, is, in the final words of Spinoza's *Ethics*, "as difficult as it is rare."

Take time to contemplate these colors, and there comes a sense of inhabiting what Kant was wont to call the dynamical, mathematical sublime, seeing the incredible way color and calculation allow us to encapsulate an entire repertoire of sound in an individual bird, an entire year of sound in a single place. This is what our crunching of big data can accomplish. But rather than simply using all this evidence to prove a scientific hunch, it is also a moment of beauty to apprehend such a picture and realize how much evidence it contains. We must feel the data as much as we dissect it.

I would like to know what makes one nightingale song better than all others, and my own hunch is that claiming it's the longest, most complex, loudest, and most outlandish cannot be the answer. Evolution is far more nuanced than that. Each nightingale species has its own specific aesthetic, and the more time we spend playing with each one, the closer we come to finding our way into a felt sense of what is best.

I have read through the science and gone out alone to play with birds and other animals, daring myself to play beyond my comfort zone, even beyond the edge of my own species, while imagining that music might bind us together with other animals in this world. Now I want to take others on that journey with me. But first I need to explore how careful listening to an entire place can make the whole city resound with a music of many creatures and parts.

# 5

## THE PLACE OF SOUND

I've spent many years investigating the individual sounds and structures of musicians beyond the pale of the human species. I've tried to learn their songs and musical worlds by striving to join in, making interspecies music that draws us humans into the art of other creatures. Nature gets much closer in the process, as there are so many more-than-human musicians to engage with. The more I try such experiments, the more I wonder how all these sounds fit together into an acoustic whole. Will the sharawaji effect be revealed when a bird in its place will be worth more than a disembodied song?

Musicians have always looked to nature for source and guidance. Lover of silence John Cage said the whole world should be appreciated as a vast musical composition. Sonic alchemist and pop producer Brian Eno wanted his own music to be a landscape, not a mere reflection thereof. In other words, to imitate nature in its manner of operation. Jazz clarinetist Sidney Bechet said, while practicing animal sounds from the balcony of his Paris apartment, "Sometimes what we call music is not the real music."

Then what is real music? How do we make sense of the whole ecology of sounds? Sometimes I've gone to scientists for advice on how to get some answers to our deeper questions about how nature works, sometimes I've just gone out to listen and play myself.

We have seen what science has to say about the individual sounds of birds. Now what of the music of entire places? Until recently there was very little science on the music of whole ecosystems, scoring the symphony of all surrounding possible sounds.

But now, in this age of big-data analysis, the understanding of acoustic interrelationships in actual places is finally possible.

We see and we hear. Vision gives us more information, images upon images that we easily accumulate. Hearing places us in the middle of things. It makes us decide exactly where we are. They complement and do not compete. And yet how much of our whole environment can we hear and make sense of by listening? I don't mean the individual messages and codes, but how it all fits. We may have only recently figured out how to analyze the acoustic whole, but we have been wanting to do it for centuries. The quality without a name, the beauty more than the sum of its parts, the ineffable sharawaji perfection no one can quite define, but which is obvious when one finds it.

The beauty of the whole is what ecology has always wanted to show to us, without dissolving the depth of specific kinds of knowledge that hold it all together. I aim to encourage total listening, or hearing the ripe presence of the whole, and our sudden central place in the middle of it. Sound is all around us and situates us in the actual world.

We seek the "perfect sound" in the novel, the sharawaji in the strange. We can design the world, but only to a certain extent. We can listen endlessly and want to know our spot. If we take it all in as sound, there emerges a "humanimal" soundscape with a place for us all.

I always try to avoid hearing the sadness of a lament when listening closely to the interconnectedness of our vanishing world. I don't really want to believe it is vanishing. With all the bad news that confronts us about nature, from rising seas and temperatures to the loss of thousands of species every year, it is important through music, art, and sheer experience to take in all the sound around us, to be better listeners so that we realize just who and what we all are. These words should not come across as empty platitudes, but take you on a journey to many places and experiences, and into the exciting research just now emerging, detailing what we've been learning about the vast ecology of sound.

Today I want to listen to *more*, to think not only of single creatures singing and making music alone. I have lived as a single musician, interacting with the sounds of other species, hoping for a kind of I-Thou relationship, one misunderstood voice to another, in some kind of decipherable interspecies mystery.

We know by now that the idea of balance in nature is something of a spiritual myth, that nature loves chaos, fights its way through to survive. Yet each part does have its place in a dynamic, messy whole that *works*. Systems have seemed too confusing and convoluted for us to speak of them too deeply, since the original heady days of systems theory in the 1950s and '60s and the earliest claims for the power of quantitative reasoning. These days, however, we are getting so much better at sifting through huge amounts of data. At least our machines are getting better at this; I don't know any people who like sifting through terabytes of data. We need machines to crunch all of this for us because we are overwhelmed by what our senses can't take in.

Science has always been a history of the interpretation of images, with experts becoming connoisseurs of ever-improving visual representations of our complex world. Sounds appear and disappear, impossible to bear in mind long enough to analyze and to count. But *pictures* of sound, in which whooshing noises like airplanes taking off or cicadas whirring become beautiful images of precise organization, can now compress entire days into a single frame.

The Ecosound Laboratory in Australia, run by Michael Towsey, is pioneering the automated categorization of individual natural sounds so that the changing soundscape of an entire day can be seen in one picture. The first image in plate 5 shows what a standard sonogram of a day's sound in the Queensland bush looks like calculated by standard computer software, while the second shows what Ecosound has managed to do with it. The upper image is opaque and shows us nothing except that sound begins at dawn and continues throughout the day until nighttime, when insect sounds take over. But with the colorized image, human examination combined with automated tools has overlaid the picture with various acoustic indices that focus on different aspects of the day's sound. This is an astonishing analysis of a vast amount of sound. On a larger scale, Towsey has been able to study eight months of single days and put forth the result in one picture (plate 6). This represents far more terabytes of listening than any human could ever go over in one or many lifetimes. Here we have stepped back in the scale of human observation. We do not see a mess but instead have line after line of movement from silence to different intensities and frequencies of sound and back again. During the spring and summer (November through March) in the southern hemisphere, there is more sound at night, more insect and amphibian activity, and the strongest dawn chorus comes in spring with that deep ultramarine blue.

What is awesome here is the sheer amount of data that we can take in at one glance. Now imagine what it would be like to listen to all that, over time, every day, twenty-four hours a day, all the while trying to make sense of something so vast. No human being's perception of this data could ever match a machine's absorption of it. The image the machine spits out is beautiful, turning all that data into something sublime.

How can we tell if one soundscape is better than another? Lars Fredriksson recommended the search for the sharawaji, but when the pioneering acoustic ecologist Almo Farina needed to invent something to measure, he decided on a number because he wanted to make his field more scientific, thereby earning it more respect.

With the help of two graduate students, Rachel Malavasi and Nadia Pieretti, Farina spent many years trying to figure out the ways birds allotted their sounds across a specific soundscape, from the moment they arrived upon migration to the time, a few weeks later, when they had paired up, nested, laid eggs, and situated themselves more firmly in the soundscape. Intuition would suggest that the forest would sound different after this time, but only the meticulous and recently developed tools of soundscape analysis have been able to turn this hunch into a scientific report.

First, Farina had to come up with some kind of statistic he could measure, to turn the sense of balance he imagined in the soundscape of well-adjusted birds into a number that could be calculated and compared. He called it the acoustic complexity index. He began with the observation that a healthy ecosystem full of singing birds in springtime offers up constantly changing living

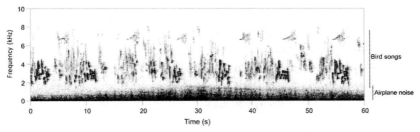

FIG. 6. A birdsong chorus above the thrum of a passing plane

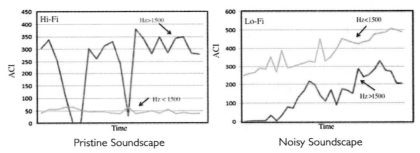

Pristine Soundscape          Noisy Soundscape

FIG. 7. The acoustic complexity index in pristine versus noisy soundscapes

sounds, whereas the *human* sounds that encroach upon such a soundscape are generally singular, such as the consistent rumbles of noise from passing airplanes, cars, trains, and machines. These sounds have an instantly recognizable quality of simplicity when compared to the busy, booming, twittering confusion of a dawn chorus of twenty-five different species of singing birds. Could this difference be boiled down into a number?

Figure 6 is a picture of a songbird chorus above the steady rumble of a passing airplane. You can immediately see a difference in the quality of sounds:

Apparently the birds are unfazed by the airplane's presence, even if it might be angering the sound recordists. It closely resembles the picture of a "pristine" soundscape diagrammed on the left-hand side of figure 7. But when more human sounds are more present and there are too many airplanes passing by, turning the soundscape into the lull of a noisy city, you get something more like the picture on the right.

The light lines in figure 7 are low-frequency sounds, under 1500 Hz. The dark lines are the higher-frequency sounds. On the right, in a noisy soundscape, the most complex sounds are the low-frequency human ones, leaving little space for high-frequency bird and insect sounds to flourish. They end up simpler and less developed. By contrast, in a pristine soundscape there is no continuous engine hum. Higher frequencies, like the joyful tunes of birds, come through loud and clear.

My innate sense is to resist the idea that data is beautiful. But look how much we can do with it today. We are perhaps at the dawn of our ability to aesthetically understand the computed world. Of course, nature will always be more interesting than that, at the same time so looming and obvious. We have better tools than ever to make sense of it, but we rarely take the time to pay attention to what at first seems to be the useless noise all around us. Could it possibly matter that this noise has been going on since the dawn of time?

The world of machines tracks us and follows us, turning us all into unique yet predictable trajectories of numbers. It is easy to ignore all that noise as well. What I would rather do is dance with all of it, make pictures out of all of it, and realize that big data is remarkably small. So tiny, in fact, that it is nearly invisible in its immensity but potentially able to help us track the beauty inherent in patterns so complicated and veiled that machines may spot the meaning in it long before we ever could. As computers give us images of what no human could ever have the patience to listen for, the ineffable depth of each moment might come into ever more dynamic focus.

What has humanity ever wanted to do but make sense out of the booming confusion of life? Are those trees outside my window random or organized? The answer is both: their patterns follow principles but will never be as perfectly simple as our ideas. Our simple images of leaf, trunk, branch, and form may be totally beside the point of emergent growth and decay.

The statistics proliferate. They are no substitute for listening and playing. We should enhance our powers of perception, seeking to capture those sounds truly emblematic of each place.

What kind of sound is it that truly takes you *there*, into a perfect representation of the actual world? In the basement of a museum in Paris, a whole world of soundscapes courses around me in the form of glowing neon lines. I am at a fabulous exhibition at the Fondation Cartier entitled "Le Grand Orchestre des Animaux," inspired by the book by Bernie Krause first published in English as *The Great Animal Orchestra*.

Krause is a peerless figure in this field because he is neither primarily a scientist, nor a researcher, nor a musical performer in one particular genre. Krause is a master listener and nature-sound recordist. He has done so many things. He replaced Pete Seeger for a few concerts by the folk superstar group the Weavers in the 1960s. He was an early sales rep offering the Moog synthesizer to the film industry and pop stars. He has long traveled the globe recording the sounds of nature, starting in the 1970s, and in the 1980s became one of the very few people to prosper by selling recordings of nature sounds, as CDs from his company Wild Sanctuary became available at Nature Company stores in malls all across America. Krause has told his story many times in beautiful books that include soundscapes and tales from his life in the field. He came up with the "niche hypothesis," which suggests that every sound-making creature has its own particular frequency in the acoustic spectrum, a sonic niche analogous to an ecological niche. Unlike most scientists, he promoted his hypothesis in the media for years without ever trying to test it. Only recently has he begun to collaborate with bioacoustics scientists such as Almo Farina to figure out what this hypothesis might mean in practice.[1] The hypothesis has become well-known in the popular press, and now he is trying to see if it can work as science. Not exactly the protocol of most scientists, who would rather propose a hypothesis, test it, and only then try to publicize it when they have data to back it up.

When the Fondation Cartier sets up one of its semiannual exhibitions, it tends to go all out. The director, Herve Chandé, was impressed that he found contemporary Congolese paintings of whole jam bands of animals getting their groove on in the rainforest, with electric guitars, amplifiers, and big PA systems: emus, giraffes, crocodiles, and gorillas playing together in a surreal Soca of bright colors. When Krause's book on soundscapes of nature came out in French, he knew that a building-wide exhibition on the subject of animal sound, animal art, creatures, color, and sound could be mounted in their beautiful Jean Nouvel–designed space in the Montparnasse district of Paris. There are contributions from great artists and photographers—Christian Sardet, Manabu Miyazaki, Pierre Modo—but I'm going to focus on the unique sound and video projection installation developed by the British team United Visual Artists in collaboration with Bernie Krause entitled "The Great Animal Orchestra," which consists of three sides of a dark room on which a video projection moves clockwise (plate 7).

It's a modified sonogram generated in real time, and occasionally the names of identified species come up in the soundscape. The whole installation runs almost eighty minutes, going through a series of soundscapes Krause has recorded around the world over many decades of travel and sound gathering. On the left there is a more abstract screen, horizontal lines of light intensifying and fading as a result of the sonic onslaught. It's like a traditional screensaver sound visualizer, but subtler and more precise. On the floor in front of the moving image is a reflecting pool, and ever so occasionally a tiny pin peeks up from the water's surface and moves the liquid in concentric rings. The curator told me this occurs at the only two moments where Krause thought that sound *might* move water, because of its frequency and intensity. The sound doesn't actually move the water, but it *could*, so the possible effect is created by mechanical means.

The message of Bernie Krause's recordings is clear. Listen to the forest after man has clear-cut it. It is not healthy; the sound is drastically reduced and compromised. It is no longer a complex soundscape but a simplified, human-tinged soundscape. We mess with a place and we take all the beauty away.

Yet when I hear these beautiful soundscapes, I start to wonder: how much of this has been found, and how much composed? Krause explains that in each case this ratio differs. Regarding the oceans, for example, he has constructed a complete soundscape of the most beautiful underwater sounds that exist, from humpback whales to the eighteen-spined sculpin fish, which sounds like a giant, low gong. Under all of it is the restful sound of crashing waves on a shore, something you never hear when listening underwater, where sound has a curiously muffled quality and it is impossible to tell where anything is coming from or how far away it is. We expect underwater sound to be booming and full of echo, so sound designers usually add that to reinforce our expectations.

We also have the expectation that when humans enter nature, they will probably mess it up, and so begins the grand narrative of loss. And while the narrative may be true, it is far from the only way to tell this story.

I for one am more interested in the sound that defies expectation. Some people find the swish of cars one after another on a wet highway to be calming and restful. Right whales approach the low drone of ships because they like the sound, despite the fact that their ability to communicate across vast ocean distances is hampered by the ships' groan, to the point where dangerous collisions have occurred.

Listen to the work of the world's best field recordists and a certain brilliance is immediately palpable. For example, many have praised the recordings of the Estonian radio producer and naturalist Fred Jüssi, the David Attenborough of his nation, chronicling the calm and specific beauty of his country in sounds, words, and images for decades. His short radio pieces, each only a few minutes in length, about the birds of Estonia were so celebrated that public buses, which often had the radio on all the time back in the Soviet days when these recordings were made, would pull off to the side of the road, turn off the engine, and take a short break so everyone could hear these brilliant and beautiful sound stories. When the program was done, the bus would continue on its route.

I asked Jüssi, now in his eighties, what makes his recordings so special. He told me it was not so much the equipment he used (though he did have the Nagra, the best portable field recorder of his day—analog, of course, made in Switzerland), but the way the programs were made:

There must be at least five years of careful listening before you can present anything in public. It was a collective work, when we did it—sound engineers and technicians also listened to our "daily catch," analyzed and edited. Every little detail is of crucial importance: levels, background noises, timbres, species.... In addition to species, also particular individuals and their talent must be taken into account when judging recordings and making choices.[2]

Jüssi spent years listening in nature before he decided to show the world what he found. He never trusted his own judgment alone but admitted that sound recording and radio should be a collective process. The best art, he seems to tell us, takes time and cooperation. He is therefore not so pleased with every Estonian nature sound release:

> You must ask: what is the aim of your recording? That guy from Saaremaa, Ivar Vinkel, his "Hour of the Nightingale" was made after only two takes. All of a sudden the CD appeared on sale; he made just two recordings and immediately went to the market—Edvard Munch worked twenty-five years on a few motifs, the scream, the kiss, you know what I mean.
>   We listened to his and my recordings, a fifteen-minute sample of each, together with the students from Estonian Academy of Arts. One of them said: one makes you restless, the other, peaceful. Vinkel just went and spoiled the delicate niche of nature-sound recordings for the sake of money.

Vinkel may have put out his disc immediately for the sake of expediency. And digital sound recorders are so much easier and more consistent. The sound might not be as subtle or warm, or maybe the birds themselves in each recording had different levels of skill. I've listened to hours of them out in the fields, but not five years' worth of continuous listening, as Jüssi says I should. It was twenty years ago that I first heard the birds in Helsinki, and I still remember that moment of surprise that the legendary bird of Shakespeare and John Clare sounded like this, something the poetry about it could never quite describe.

Nightingale, the unattainable—the absent, the call of the Old

World missing from the New. They challenge our musical expectations. If you are right next to one hidden in a dark Berlin bush, they are loud enough to peak your recorders, as they have evolved to be heard from quite far away. In all my recordings I tweak their frequencies, using equalization (commonly shorthanded as EQ). You have to pull out those 2.8 and 3.4K frequencies—they are just too strong and jarring to the human ear—though I know full well they must titillate nightingale ears, or at the very least tickle specific neurons in their brain. They know what it is they have to know, which sounds do and don't matter.

Jüssi's bird seems to move a little left to right on his celebrated recording, likely something one of his consulting audio engineers added. It seems live and real on the headphones but somewhat contrived. I imagine on properly placed speakers it adds a little living movement to the sound of the bird, so he seems more like a creature than a lone sound. All this technical info should be secret to the listener. Who wants to know any of this stuff when taking in something so beautiful? The recording is a conduit to an idea of a sound that may be more than the sound itself can ever be. I was always intrigued that cardinals, those most astonishing of the common birds round here in the Hudson Valley, have been found to be more impressed by synthesized versions of their own sound than by the real thing. Similarly, in the laboratory Ofer Tchernichovski and I were able to get a female zebra finch to be more excited by remixed versions of her mate's song, repeating over and over again far faster than a real bird could ever do. Like humans, these birds sometimes prefer simulacra that are larger and "purer" than life itself.

Maybe that's what the best recordings are supposed to do: *not* document the soundscapes of the natural world but use the possibilities of art to complete what nature has started. That old argument about *techne* that Aristotle imposed upon us thousands of years ago may be the root of the whole hubris of Western civilization. We want the most beautiful, transformative sound of the world, not necessarily what is real. Is humankind simply incapable of bearing that much reality? Is there truth in natural soundscape recording? We want photography to be creative but not hyper-creative. Giant, shiny, superrealistic images should trouble us as much as any other human lies about the world around us.

I think of these things as I walk to the riverbank not far from my house this morning, maybe the first autumn morning of this un-usually warm season. In front of me is perhaps the most beautiful view of Little Stony Point Beach that I have ever seen. There is a fog ahead over the bay to the north, beyond Storm King Mountain. It is not receding but approaching, very unusual for a morning scene. The Catskill mountains in the distance are totally invisible; even the water surface cannot be seen as it blurs into the clouds; it is as if the water itself is lifting from the Hudson on up to the heavens. I don't want to take a picture of it because the mood is so amazing, so instead I start to listen.

There is an unusual breeze that leads to small, consistent waves on the river's edge, like waves on a tiny sea—a sound one rarely hears here, in this way. It's a perfect rhythmic ground for every-thing else I hear. The gusts of wind come in and out of the trees, producing washes and noise whistles that come and go unevenly. A blue jay makes a onetime cry. I try to isolate the elements, follow their differences: rhythmic waves, enharmonic wind, occasional animals noticing their place. This I would have liked to record, then to listen and then to decide if it is any *good*.

Some of my artist friends laugh at my fixation with the good and bad, the better and the worse. Too much art I come across could be improved if artists were more interested in the difference be-tween better and worse. They tend to say this is an outdated binary carried over from the last century. But without it what are you left with—the popular or the unpopular? The brand-name artist you remember or the one with the most famous friends? That's no way to touch people with beauty.

Fred Jüssi played me one of his favorite recordings, a twenty-seven-minute track of the shores of the Baltic Sea one morning in Rohuneeme, Estonia, in the spring of 2005. "Everyone who listens to this track feels a total sense of calm," he tells me. "They cannot turn it off." I feel the same thing, the overlapping winds, occasional seabirds, crickets in the reeds. "But you know what?" he smiles. "We only had seven minutes of audio. The engineers cut and pasted it together so it doesn't sound like anything reappears."

Now I smile. The favorite work of this great audio documentarian is actually a composition, a careful construction made to sound not like a composition but like the reality of found nature. The best listeners know how to construct beautiful works that sound better than what you can find out there in the world. Is this the true power of media?

I went to visit Lang Elliott, arguably America's greatest nature-sound recordist. He is among the few deep listeners of nature who have a highly developed sense of aesthetics, and he is currently interested in figuring out why certain recordings sound better than others, are the most immersive, and seem to promote a healthful state of mind.

Elliott began his nature recording career while working for the Cornell Laboratory of Ornithology. After obtaining a grant from the National Library Services for the Blind and Physically Handicapped, Elliott produced "A Bird Song Tutor" to help blind nature enthusiasts visualize their natural surroundings. Shortly thereafter Elliott left Cornell and set his sights on producing narrated audio guides to bird and frog sounds in the tradition of Arthur A. Allen and Peter Paul Kellogg, the original founders of the Lab.

Like the wanderer in *Zen and the Art of Motorcycle Maintenance*, Elliott had to figure out what quality means. After twenty-five years on the job, he has a simple and clear answer: a quality recording is one that takes you *there*. It's a sound that situates you in a wild and natural place. Unlike some recordists, he doesn't think these places are all that hard to find, although periods of true quietude are elusive and often brief. "Near where we live, there is wildness, there is relative silence and peacefulness, augmented by the fact that we tend to filter-out extraneous background noise when we're in natural settings. Maybe it's not all that many minutes before a car goes by or an airplane flies overhead, but the sounds of nature are widely accessible. So we need to capture them in as pristine a manner as possible and present them so that more people will learn how to listen to and to value them."

Elliott favors a modified binaural approach—that is, two microphones mounted in a special housing that re-creates the way we humans equipped with two symmetrical ears listen to the world. Many nature recordists swear by the contraption that Lang uses, which is technically referred to as SASS (Stereo Ambient Sampling System).

Binaural recording technology has been around for decades but was not taken so seriously, because people in general did not listen to music with headphones. Now that most of us do, it is a method whose time may at last have arrived. Elliott improved his setup by installing better microphones, and he has been carrying his SASS with him for many years during travels to wild areas throughout North America. His recordings work best when heard with good headphones, but the technique translates to stereo speakers as well:

> I love recording binaurally and then enjoying my recordings later in my studio or while at home. It may seem like it would be easy to capture a great soundscape, but it's not. Truly wonderful soundscapes are the result of careful choices and a bit of luck. When and where to record? Under what weather conditions? Where exactly should the microphone be placed? If one chooses correctly and the force happens to be with you, a masterpiece may come into being.

Technology is one thing; attitude is another. He has always been searching for the most spacious and captivating sound. "The answer lies in aesthetics," smiles Elliott, who has just come through a tough fight with throat cancer. "My voice has now recovered, I'm ready again to talk about these things. I will be podcasting, I want to share my stories about how to listen to nature."

Sound is the best human sense to situate ourselves in a place. We cannot close our ears, so just as sight gives information, sound sets us in the middle of a world. Yet, when we think of the representation of nature, we have much more to say about the visual. What makes a good landscape photograph? Is it the right framing of the image? A well-balanced composition? This is analogous to finding the right microphones, the right mix of sounds near and far, of background versus foreground.

There is a kind of nature photography where everything jells, a mixture of beautiful subtlety and the precision of technology. A precise shot of an eagle catching a fish, the kind of thing only professionals used to be able to grab, is now attainable by consumers' ability to shoot thousands of images with precision lenses. More of us can have the technology, but it means little without seeing the potential for beauty.

Is soundscape recording, or soundscape listening, all that much like nature photography? Nature photography sometimes seems easier than other kinds of art photography, because nature almost always looks good. We know there is something right about it, a panacea to our complicated, human-dominated lives. It is either beautiful or boring, depending on your mood or who you are. The technology we now have makes it astonishing, glistening, 4K, 8K, 27 megapixels of exactitude. Blow it up and display it on a high-resolution screen, print it out on glossy, indestructible metal, or shrink it to fit on a phone screen and spread it virally around the world.

What is quality? What is the best nature photograph or nature sound? We imagine we have seen or heard it all, but we have seen and heard hardly anything because we don't know how to pay attention. Sometimes attention begins with an idea. Like that unusual *National Geographic* cover photo of Yosemite that tries to show night and day in the same image. The photographer tries to show all the best shadows and light on each part of Half Dome and the whole valley by stitching together hundreds of images of the same scene, taking the best coloration and shading from each, from midnight through sunset, like a time-lapse single image of one spectacular site. Stephen Wilkes is trying to make a single image show the procession of time over a landscape where no things move.[3] It is the visual equivalent of an hour-long recording that might claim to be a day in the life of a rainforest, where all the sounds you need to hear are heard, from a rainstorm to a coqui to a screaming piha to a jaguar, even though in any single hour you would never hear all these things.

The result of media will never have too much to do in duration, intensity, and structure with what you might really find in the wild. As Elliott would have it, the work must make you feel what it is like to be *out there*. It will be beautiful and perfect enough to transport you into the natural world itself.

Critics have made fun of this honorable desire for centuries. Remember Diderot's famous essay on the Salon of 1767, where he is supposed to review an exhibition of landscape painting by Vernet and refuses to go into the gallery, instead heading straight for the forests that the paintings purport to represent.[4] Of course this is a better place to be! Perhaps our landscape-inspired works exist solely to remind us of the superiority of nature to art. We are all children of artifice and can never live in the middle of nature, no matter how much we pretend. We need to be reminded of what is great about this world whence we came.

John Muir takes a tour group of artists to Tuolomne Meadows, where they seek just the right view to paint, the perfect place to set up their easels. He gets so fed up with this crew that he disappears for a few days into the very landscape they are trying to paint. He gets soaked while sleeping through a furious storm, dries out, is enveloped by the sun, and charges down the mountain energized and more alive than ever, where he finds the artists worried, nervous, and uncomfortable, but talking about the beautiful storm clouds they were trying to commit to canvas. "I know," smiles Muir. "I was there, in the midst of it." A feeling he will never forget.

Nature is impossible to represent in art. This is why *Grizzly Man* is my favorite nature film. In it Werner Herzog treats the subject of Timothy Treadwell, a man so obsessed with bears that he ends up being be eaten by one. With all the time Treadwell spent among bears, he got far closer than Herzog, master filmmaker and philosopher of the abyss, would ever want to go. Yet the film Treadwell would have made, had he survived, would not have been appealing to anyone but himself. He had shot incredible footage, but he was not a film director. The world is lucky that Herzog got hold of that material and decided to make such a perfect film. It shows the ultimate folly of humans claiming they can understand anything about the world of animals or the way they perceive the world. We are always only ourselves.

In the most moving scene in *Grizzly Man*, Herzog listens to the final audio recording of the grizzly bear eating Treadwell alive. You see his solemn expression as he hears the terrible screams, silent to the viewers' ears. "No one," he announces to us from the screen, "should ever hear this tape. It must be destroyed." If nature is as terrible and dangerous as we fear, this ultimate terror should never be repeated for our entertainment. It offers a deeper truth than that.

You cannot forget all the art you have seen or have heard as you strive to make your own. We are only as original as our history. Take everything in and you will learn how to fit in, as a human animal in nature, and as an artist in the world of art. It's the same for image as it is for sound.

Composing is only a bit like painting, sound recording only somewhat resembles filmmaking. To make art out of any of these media may take years of attention and learning, or it may take a certain knack or talent that then must be recognized. Lang Elliott learned early on that he had an ability to tell when a recording was good. He found this talent drew him deeper into the field than the need to answer questions via scientific method. Like so many great lovers of nature, he never finished his dissertation. He had too much else to do.

Since an accident that occurred in his youth, Elliott has not been able to hear any sound above 3000 hertz, near the top of a piano's range. He misses much of what birds and insects are singing. So how can he make such fabulous recordings? Maybe this limitation made him even more determined to listen well. He developed the first truly advanced hearing aid for birdwatchers, the SongFinder, which shifted the higher frequencies down in precise ways to enable the many birdwatchers who don't hear as well as they used to notice what's out there.[5] Through years of practice he's been able to compensate, to know what needs to be heard. A great composer of soundscape recordings listens to so much in the world, and has the ability to know what is worth bringing home.

A sound without a context is just a raw, technical example, a species name to file away on a list or a song to add to the names you think you can pair with a sound. That's no way to put forth beauty. A perfect image of a bird on a white background might be good for identification purposes, but not for any understanding of life:

> One morning, after it had rained, I recorded a hermit thrush at dawn along with water dripping from the trees. There is one focus sound object, the thrush singing, accompanied by countless drips … thousands and thousands of ever-changing sound objects. The drips spread out in space to create a three dimensional scene or envelope within which the sparkling song of the thrush is periodically heard. I found it in nature, it's the real thing. I wouldn't be impressed if someone created or constructed this.

"It's one thing to recognize the beauty of the hermit thrush," he explains, clearing his throat with a smile. I get the sense he's wanted to talk about this for some time now, but that no one has asked. I have read and written extensively about what people have said about this most melodious of North American birds. Flute-like, whortling, pentatonic or not, as T.S. Eliot wrote in *The Waste Land*, "Drop drop drip drop drop drop drop." Most noticeable to me is the thrush sounding musical in a nonhuman way, but musical all the same. We *get* it, unlike so much of the natural world, which eludes us.

Elliott doesn't want to analyze the bird's song; he wants to present it in an aesthetically pleasing manner. He needs to find in nature the perfect placement of the song, the instant composition. If the bird is the foreground, he needs a background: "When the sound really takes you there, it is almost like the moment in Zen meditation when you let that journey be your story. Some people think the more you know, the deeper the experience, but too much knowledge of a bird or other sound object can take you away from a bull-bodied experience of place, of the natural surroundings."

So many of Elliott's most beautiful soundscape recordings make use of this simple aesthetic approach. Snowy tree cricket rhythms interpolated above the gentle waves lapping the shores of Lake Ontario. Cricket frogs overlapping with pine sawyer beetles chewing rotting logs. So often there is an overlap of two main sounds in an Elliott recording, and even in his dawn chorus recordings, at these moments so often overwhelming with sounds in the height of spring, Elliott finds clarity, a kind of chamber music, in "Summer Frogs," as in many of his other soundscape works that can be found on his website. These recordings are simply brilliant, clear, and present.

Elliott never constructs these soundscapes in the studio, which is a valid way to make music or nature sound art, whatever you wish to call it. He is not a purist, for he does sometimes use noise reduction, equalization, and some of the subtle editing tools of the modern audio production world. Most important for him, however, is to find real natural moments that have a kind of balance, harmony, and clarity and then isolate these from the many extraneous hours of field recording it took him to get there. It's akin to finding the best shots out of thousands of photographs and presenting them as real moments from the natural world.

There are countless immersive natural soundscapes and transects that move through the arc of every day and night. Most of them are never recorded or noticed, like all those trees that fall in the forests of our unremembered and undocumented world. Have we not recorded enough hermit thrush songs and frog choruses to satisfy anyone's curiosity? Of course not. There are many more great photographers than great sound recordists, and we still need to take more pictures because we always aim for a personal connection to the beauty.

Elliott is quite humble but does admit one thing: he has a knack for it. He somehow has learned to record beautiful sounds in nature, although his learning curve has been long. Upon leaving Cornell, he focused on getting clean and clear recordings using a parabolic reflector microphone, which effectively removes a bird's song from its surroundings. Only later did he begin to focus on gathering beautifully immersive and expansive binaural recordings that truly capture the essence of place and time. There is no doubt that Elliott, much like Fred Jüssi, is an artist in a medium that he has made his own.

It is difficult to explain what it is to be a good listener to nature. It doesn't mean you get more facts out of the process or you write more numbers down; it is more about recognizing a beautiful confluence of sounds and being able to represent such moments in some way that you can take back to fellow humans not fortunate enough to have been there, and to present the possibilities of rare beauty that lie all around us yet go unnoticed.

A beautiful image can be made of the most mundane of moments. The same holds true for a soundscape recording. The wild is always a grand subject for photography or sonography, but it can be *too* easy a subject. Getting the right image or sound means noticing the gripping and the important before choosing the right technology. Cameras and recorders have gotten good enough that any one of us can take a crack at it. Just practice listening and watching. Jüssi recommends you practice five years before you reach out to the public. While that does show a noble humility, even *National Geographic* knows its readers can sometimes get a better shot by happenstance than their professionals can after hours of calculated framing and waiting. Our desire to freeze the tumult of the world can and does work like that.

Talking and writing about aesthetics are never easy, but that doesn't mean we shouldn't do it. We *must* do it, unless the mediocrity of ubiquitous media is enough for us. Block the boring, and only welcome the extraordinary. Train yourself to know where and when it appears.

Sometimes Elliott falls into the same trap when he tries to explain why his audioscapes sound so good. They're relaxing, they'll help you fall asleep, calm you down. He now wants to develop "sound medicinals" to help us feel and function better. He shouldn't need to do that. The best music already makes us feel more alive, and if it comes from nature it should make us love the natural world even more, feel more connected to the universe, never lonely and self-contained. It does not have to justify itself with a practical end, although maybe my view is increasingly in the minority in an age where every moment of the day must further our pragmatic goals of becoming better, faster, and stronger. Not for me. I just want to be surprised by beauty and to teach people that this beauty requires time and attention, an about-face from all those practical goals oriented toward some arbitrary ideal of success. Spend time with the resounding natural world that envelops us. Listen closely and love it all the more.

# 6

## CALLED MOST BEAUTIFUL

A few years ago someone decided a contest should be set up to find the most beautiful sound in the world. Possibly a pointless task, but people have gotten so interested in rating everything. I can imagine John Cage being one of the judges, saying, "How could we decide? No sound is more important than any other." We only need to temper ourselves, and anything we choose to fixate on can be heard as beautiful. Sounds are sounds. After Zen, nothing ever changes, but maybe our ears burn a bit brighter, our bodies suspended ever so slightly above the ground.

Most interesting, though, is that the winner was a recording of a forest somewhere in Borneo by Marc Anderson of Sydney, Australia. This natural soundscape is like a complete piece of music, with rhythm, texture, different instruments in different places, not so much any beginning or end but a resounding *middle*, a dynamic moving texture where each sound seems to have its purpose, balance, and order. The organization becomes clear when you print out a sonogram (see plate 8), and it is impressive how the different species of frogs, insects, and possibly one moaning bird all take their places among the sonic frequencies, demonstrating Bernie Krause's "niche hypothesis" of each animal finding its own place in the thick of it.

These sonograms—easy to produce on computers or phones with software such as AmadeusPro, Sonogram, Sonic Visualizer, Raven, or Sound Analysis Pro—map frequency against time and enable us to grasp the structure or confusion inside complex samples of any sound, natural or otherwise. I have been throwing sonograms at you without explaining precisely how to read them, opting for an immersion approach in lieu of any systematic explanation. (I use Amadeus because it seems to produce the most aesthetically pleasing results, where a little tweaking can reveal the most interesting aspects of a sound that we sometimes cannot hear.) Know at least that a horizontal line is a single held tone, like a clear note on a flute or a nightingale's whistle. A vertical line is a click or a slap. Parallel horizontal lines, straight or curved, indicate harmonics or sounds with a rich tonal quality, such as those of a stringed instrument or foghorn. Fuzzy, noisy, complex, or beautiful sound images denote a hard-to-describe sound, such as the whirr of a cicada or the sexy buzz of a nightingale.

I love the sense of rhythm, whoops, moving clouds, and one moaning tone this sonogram reveals. You can see and hear why some would consider this the most beautiful natural sound in the world. It is a composition of many natural sounds, a fragment of an ambient symphony. It's surprising that nature could offer us such emergent music, with no one in charge.

That is an environmental beauty, and I can see why everyone voted for it. The voice of the public cannot be ignored, and maybe picking the most beautiful of the mostest really is a popularity contest. But is it really more beautiful than the awesome trumpeting of a herd of elk in autumn heat? In Elliott's magnificent recording, amazingly from Pennsylvania, where there is just one elk reserve, we hear soloists vying for attention in the mix, the mammal world's analog to the thrust and parry of a park full of nightingales. The section shown in figure 8 is about two minutes in.

The beautifully curved, lilting harmonics; the wavering, long cries no tempered notation could easily grasp; these single roars overlapping with passion and depth: surely there is a beauty here that a mass of tiny insects cannot approach? The comparison is nearly impossible, of course. Although in Elliott's elk recording one immediately hears the brilliance of his work, the perfect selection of a fragment of nature in Anderson's winning recording is fabulously composed.

The beauty of the singular, the beauty of the whole. A soloist, a duet, a symphony or band. The magnificence of an individual

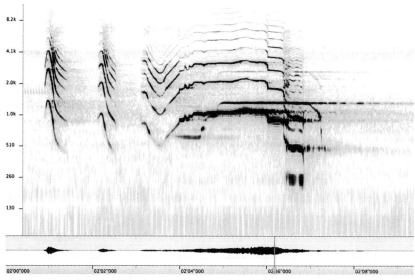

FIG. 8. Bugling elk in Pennsylvania, recorded by Lang Elliott

or the convolution of the whole. These are different ways to form musicality in a natural moment, something to seek out and capture in a recording so all of us can hear and be transported back to the source.

Transportation is the key for Elliott, the sound that makes us feel we are there. Not for any ethos, not for any comment on the state of the planet or a lament for human rapacity in the destruction and elimination of such moments, but the wondrous ability of sound to make us feel we are somewhere else, rare and beautiful, either far away or right around us, in each case often overlooked:

There are different kinds of recordings. Close-up recordings of specific species, so loud and clean, these are impressive, but very quickly you tire of listening to them. There isn't much that is *healing* here, you soon turn it down or turn it off. There are also recordings that are pitched as environmental yet have so much going on, they are too much of a cacophonous barrage, disturbing at some level. The kind I like is this third category where there are fewer elements, more tasteful, not cacophonous, that don't wangle your ear. They are engaging at some level but not the normal way, they are something you can sink into. They are Zenlike, somehow healing. My Zen practice has certainly informed my recording practice.

You may know what it is or have no idea at all, but sound can pull you right into its shape and form. Elliott wants to transport us; Krause wants guilt. Gordon Hempton wants fear for the loss of the possibility of silence. His "One Square Inch of Silence" project is especially poignant because of its locale. Hempton has lived on the Olympic Peninsula for years, and although he has traveled the world in search of the most beautiful and pristine soundscapes, he takes pride in his one local spot, deep in the Hoh Rainforest, which he says is one of only twelve places in the entire continental United States where you can hear nature unmitigated by the intrusion of a human sound for up to fifteen minutes. That's how prevalent our noise encroachments have become.

Yet we humans have always sought out the noise as much as we have whined about it. We might not be the only species to seek out clamor. I was amazed to come across a mockingbird deep into soloing away at the edge of the Brooklyn Botanic Garden, soaring above the din but somehow in tandem with it. In this lush, planned, narrow triangle with roads on all sides, under the busy flight path down to Kennedy Airport, there is no chance for silence. What a laboratory for the study of noise! There is quite a rumble above which the singer must chime. But he *is* above, his tones pitched high enough that perhaps the noise doesn't matter to him. Or maybe he's adjusted his tune to the situation. Numerous studies have, in fact, suggested that city birds sing higher, louder, and even faster to make themselves heard in congested human environments.[1] Birds don't give up but adapt. It's a booming, buzzing confusion, but they find a way to break through it. Brooklyn, Brooklyn, take me in. As much as we fear noise, we can work with it. Even the mockingbird seems to do it. In figure 9 he is duetting with a police siren.

Hard to say who is copying whom. Who should be outraged, who is playing the field? Mockingbirds make use of only some of

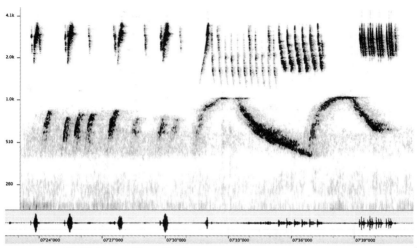

FIG. 9. A mockingbird duets with a police siren

the possible sounds around them. A precise interior sense of aesthetics helps them decide. They sing to defend and to attract, but also to be truly themselves, to be known and heard above all the fray. They do not flee to the wilderness but make their home among us. The planet warms, they move ever northward, singing their songs as the climate changes ...

I trust the inadvertent edge that isn't quite what we expect. That's why I listen everywhere for the *boori* sound. We thrive on the unexpected, the quirk of surprise. The best jazz musicians to me are impossible to emulate, difficult to diagram. Not the encyclopedic Parkers or Coltranes, but the enigmatic *boori* types: Thelonious Monk, Ornette Coleman, or better yet Dewey Redman, Charlie Haden, Wayne Shorter—those masters who've *got* it, that vexing quality that cannot be measured or written down. They have no method, intractability, no repeatable results, no absolute confidence or trust, and certainly not probability. Art, not science.

Nor am I impressed by relentless predictability; the nightingale needs this one odd sound to surprise us. Shelley's woodman wanted to vex the nightingale, while the nightingale is the one who endlessly vexes us by making sure his song always has something we can't predict. The revolution will not be simplified! Long sing the bird! He remains a beautiful puzzle, his song eliding, always just out of reach yet close enough to our music that we will always reach for him.

Even Goethe got in on the act, noting in *Elective Affinities*: "In many of its tones the nightingale is only a bird; then it rises up above its class, and seems as if it would teach every feathered creature what singing really is." I guess an elective affinity is someone we choose to become obsessed with, since that august book is all about love. Or is that metaphor all about chemical reactions? Maybe we do not decide which bird we will grow to love. Without love, life is only a shadow of itself. Goethe again:

> A life without love, without the presence of the beloved, is nothing but a mere magic-lantern show. We draw out slide after slide, swiftly tiring of each, and pushing it back to make haste to the next. Even what we know to be good and important hangs but wearily together; every step is an end, and a fresh beginning.

I don't think he means it's a series of PowerPoint slides doomed to endlessly repeat themselves. These ultimate birdsongs are never only data to analyze and reanalyze. They are essential and beautiful tones that will always need to be sung.

The Brooklyn mockingbird keeps singing in broad daylight inside the hullabaloo. Nothing seems to stop him; he's tough, that bird, although not ultimately so. After singing for an hour or more, he is at last defeated by the noise of a helicopter. This great rotorbeast approaches the beginnings of his huge mechanical song with some glimmers of rhythm, laying a steady beat underneath the endlessly mocking bird.

There are no nightingales in Brooklyn. No American bird is quite like the nightingale for singing in your face, accepting the challenge of immediate and strange sounds. Nightingales also transcend the tumult of their context, and Treptower is far quieter than Prospect Park. The grumble could be cars swishing by on the *allées* or the multispecies twitter of dawn that tends to silence the nocturnal birds. They operate in a world where the individual must peal through the soundscape. That's why they attract the solo human musician. Maybe only when our own individual sound gets lost in the whole of the soundscape does the whole of an ecology truly resound.


The most beautiful sound is something that makes you weep or sigh with the fact that this world is warm, inviting, and amazing, and has a purity and rightness that exceeds our worries and uncertainties. The confluence of a lone bird's song and an ultimately human noise may also reach for the sharawaji. That beautiful sound of Borneo still evokes an electronic composition. It has a sense of tone, harmony, rhythm, and form. I could hear it resound over and over again for hours. I think about what I could add to it to use it as a musical background or inspiration, but there's no space for addition. It sounds so complete as is. There's even an occasional pure tone at the pitch of 750Hz (F♯5) that already sounds like a saxophone put into the mix.

I'm never surprised when I hear a fragment of nature that sounds like music, and I'm always looking for more such soundscapes to lure in the uninitiated, to convince people that there is something to this wild idea. At first I scoffed at the very notion of such a contest, but when I heard the sound it grabbed me ... they just might be right. Perhaps this *is* the world's most beautiful sound. At least the most musical wild soundscape. It sounds like a forest composition, perfectly laid out and organized. No surprise that the graphic image of this music, the sonogram so easily spewed out by my computer, looks so formally designed. Who says animals can't sync to a beat? This armada of creatures resounding among themselves creates a more fluid, living jungle beat than any humans can know. We get it into it, though, and with time this music makes us work. This weekend I will take a fragment of this most beautiful soundscape and use it as a guiding background for a performance I'll do on Saturday. Every few weeks I do one of these performances, and have been for years. It starts to seem like they are all the same, like I am one of those singing birds who just has his one song, his only shtick, who repeats incessantly whenever asked.

For a while this situation was bothering me, but I was heartened to read that Karlheinz Stockhausen said he didn't think composers should have works, just one giant *Lebenswerk* that takes your whole life to complete. I suppose that is really what I am up to. So few people actually hear my music, why not just play the same concert over and over again? Even the world's biggest music stars do that because the people really only want to hear their biggest hits. How much work does each of us need to complete to feel accomplished? Perhaps I am a one-idea person: listen to the beautiful sounds of the world and celebrate them. Figure out what you can add to them. Otherwise there is no reason to make music, because the world is so beautiful without us.

Maybe it's the same message I am offering to imagine my music is ecological: don't ask nature to follow your rhythm, like some great white parrot bobbing to the Backstreet Boys,[2] but follow the wafts of rhythm in that unlikely amphibian williwaw. You will become larger than yourself and come to love the real world out there all the more.

Would Messiaen have agreed that the Borneo frog fest was the most beautiful sound in the world? Doubtful, as he preferred individual avian virtuosos, with their superhuman leaps and bounds. I know Bartók would have loved it, his *Out of Doors* suite being based on the night hoots and stridulations of insects and other inscrutable creatures of the dark. Then again, *Out of Doors* was a piano piece. Would he accept such invisible bleaters as musicians in their own right? Were he alive today, I'm sure he would, because today's composers are accustomed to fumbling incessantly with the accessible beauties of every extractable sound.

Some people are impressed when machines can make music automatically that might fool us into thinking a musical intelligence is at work behind the results. I am more impressed when nature reveals music of this kind, since in nature there is no designer who has formed sonic tools in his own image. The sounds have simply evolved to find their place, and all creatures around them can hear and can use their beauty.

Patterns evolve; symmetries expand into asymmetries that offer a perfection massaged over millions of years of steady change. There is a rightness in such sounds that no human hubris can ever hope to achieve. Yet I always wish to join in with such music. Nature lovers and skeptics alike sometimes tell me I have no business being there. Maybe *no* human has any business being there. The most beautiful sound is a perfect sound, it has no need for interlopers. And yet we interlopers want to benefit from our association with nature, our ecologizing of art. By putting forth true natural sounds in our palette, we want to be more real, more necessary, closer to the Earth, no matter how much technology invokes. Either a shakuhachi or a laptop can reveal the distant sound of deer calling to one another in the night. Which one is more real depends on what reality you aspire to, either the direct pull of breath and air into sound, or the direct transformation of the sincerity of the noise. No sound is ever safe from being considered music.

"I will never say that I'm a good listener," said Gordon Hempton, the Sound Tracker, another of the world's greatest field recordists. "Thinking that I was a good listener was one thing that kept me from being a good listener."[3] I am intrigued by his lifelong search for silence. He looks for inner silence, which he says is a reverence for life and the good that we carry wherever we go. But he also looks for outer silence, a naturally quiet place where human sounds do not intrude, and this is something he can hardly find. Crisscrossing the whole United States, he has identified only twelve places where nature can resound without being interrupted by a human noise less than once every several minutes. When it comes to sound, there is no longer any nature that sings without us intruding on the mix.

John Muir, 140 years ago, listened everywhere he went in the Sierras and beyond: "The profound bass of the naked branches and boles booming like waterfalls; the quick tense vibrations of the pine needles, now rising to a shrill, whistling hiss, now falling to a silky murmur, the rustling of laurel groves in the dells, and the keen metallic click of leaf and leaf.... The air is music the wind forsakes. All things move in music and write it."[4] Muir wrote these lines in an age before sound could be recorded. We needed such words to believe that the Earth as a whole is at least a musical being. He heard sound as one more piece of evidence of the beauty nature calls out, if not for us then at least to us, ready for our attention and reverence. For Hempton, more than a century later, silence is both a practical and a spiritual goal; it is the principle that guides all his work and his message.

By his two kinds of silences he never means no sound, just no unwanted sound, and no mindless machine-made human sound, the noise that we cannot escape. Yet more than his ethos, his life journey and choices are what intrigues me most about Hempton. He was engaged to a deaf woman for several years—did he find in her someone who was at peace with silence? He has chosen to live as close as he can to the quietest of all American national parks, the Olympic, with its Ho Rainforest, where Hempton has identified his own personal shrine, the one square inch of silence where finally the encroachments of noisy man can be fought off. His recordings are astonishingly beautiful and precise, clearly composed and constructed, yet he claims he does not edit them. Why does he try to say that when they are so clearly works of art and not documents of nature? I do not know—but it may be akin to those photographers who use Photoshop to correct color but would never move an object into an image that was not originally there.

I am wrenched by the tragedy that Hempton lost his hearing not once but twice, and am inspired by the fact that he is partially able to hear again but now cannot make such exemplary nature sound recordings without the help of someone whose ears are clearer. Seeking silence, he has been given silence, but he will not rest because that silence does not really exist in the actual world out there. We still must reckon with the omnipresence of noise.

Hempton's latest book is a self-published pdf called *Earth Is a Solar-Powered Jukebox*. I can see why Hempton elected to publish it himself; this way he could do everything exactly the way he wanted—color photos, links to online audio material, no publisher telling him what he could or could not do. Unlike his previous memoiristic *One Square Inch of Silence*, this recent tome is essentially a how-to manual for working with environmental sound, aimed at acousmatic composers and especially professional sound designers for games, podcasts, and films.

His most interesting chapter is one on how to record what Hempton calls *quietudes*: natural ambiences at the very lower limit of human hearing, soundscapes that are almost silent but not actually silent, the real, beautiful quiet that is so close to silence but not empty, like the engineer's anechoic chamber immortalized by John Cage.

Hempton is famous for saying that there are really only twelve truly quiet places in America (and none in Europe), where one could go just fifteen minutes without hearing an interrupting human sound. One is the bottom of the Haleakala Crater on the busy island of Maui. Go ten thousand feet up and then walk one thousand feet down, and you are in the middle of a strange tropical desert. He calls it the quietest place on Earth, where the sound level "measures in negative decibels and feels absolutely silent at first listen, but when amplified just 20 dBA we can actually hear the mantra-like beating of Pacific sound waves lapping over the ten-thousand-foot-high rim of the volcano." Such barely vibrating soundscapes are quite tricky to capture:

> Once I've chosen a location I commonly experience the thought: "Nothing is happening." I've learned that this once-disappointing observation is actually great news. . . . Early one morning, from where I sat, headphones on, the whole world was asleep. Listening to my recording now, my eyes have filled with tears as I recall that glorious morning and the revelation of how peaceful a natural place can sound in sustained anticipation of the rising sun. It is the presence of everything. Quiet is profoundly quieting.

And of course such quiet is profoundly rare ... I wonder if noise can also be quieting? The spirit and aesthetic of Krause, Hempton, and even Elliott are not my choice. I delve into the sound of nature because I know how much it can become like the sound of humanity. The waft of waves can resemble swishing cars on an expressway. The rumble of thunder can resemble an approaching train. Insects sound like electronic music, whales moan like the din of approaching ships. The message of the sound conservationists is right: we are losing the purity of nature. We are losing species, habitats, animals rare and common; all are going down as we continue to trash the planet. The news is bleak, there seems no way out save all-out war on the wastefulness of all our human ways. I do not doubt the accuracy of this pessimism.

Sound is fluid and can fool you. In its acousmatic purity it can entrance us when we have no idea what it is, human or animal, natural or artificial. It is fascinating to hear David Toop's almost anti–field recording of his walk through a Chinese market where both real insects and insect toys are for sale, and along with the rush of motor scooters and heckling salesmen we can hardly tell what we are listening to, the animal or the machine. Only one thing is clear: that this sound world through which he walks is relentlessly alive with a mess of species doing their best to be heard.

I am listening to Gordon Hempton's wonderful track composed out of sounds from his favorite soundscape in all the world, a piece that accompanies his book *One Square Inch of Silence*, which refers to an exact place a few miles off the trail in Olympic National Park. The piece begins with howling coyotes. Is it natural or artificial reverb surrounding their melisma? Then comes a beautiful rhythm of raindrops on leaves, coolly cyclical, an easy beat to sample and work with, like footsteps or the clicking of whales—sounds that tempt us hunters of rhythm. The piece is a series of solaces, silent spaces between offerings, like individual songs of nightingales with space for response or reflection in between. Now Bernie Krause notes that "humans have been found in the Hoh Rainforest long before Hempton was born. Does his white sensibility suddenly matter more in the arc of human events?"[5] A valid point. We are the people today to value silence so much, and yet as Krause has often eloquently written, there is no complete silence in nature. Let it be noted that these great sound recordists don't shy away from scrutinizing each other closely.

Silences, be they inner or outer: what are they really? Certainly not absences of sound, but spaces in the mind, moments to reflect within. Most who've been inside anechoic chambers, artificial rooms that have absolutely no sound or resonance, are unhappy with the experience. John Cage was that rare exception who spoke of the experience most positively, saying he heard the whirr of his brain and the beat of his heart. Others feel queasiness, some sense of sonic space to be missing. For better or worse, we cannot stomach true silence. We are at home in noise, the booming, buzzing confusion that William James defined as the stream of consciousness that keeps us alive. Isn't meditation supposed to slow all that down as we clear our minds? "Not really," said Leonard Cohen. He ought to know, after spending forty years sweeping at the feet of his Zen master on Mount Baldy. "Meditation is best at stopping us from whining."[6]

And whining often seems the natural reaction to despair at all that's wrong with our human take on the natural world. We are extracting everything from it, in the process losing the beauty that has been all around us for millennia. I cannot dispute that, but I refuse to be stuck there all the same. Let's celebrate the mystery we can encounter with those tough sounds in nature that survive, those enduring birds that can sing through the night.

In some of its range, notably the English countryside, the nightingale's sweet song is hardly heard anymore, and we should do what we can to save it. But in other places this great bird's song is expanding, ascending to the tops of trees in great numbers, and we may now take the time to celebrate its tenacity in the face of all and everything toward deepening our listening.

I wonder about these things as I walk from my house down the steep autumn street to the river. This is far from the best time of year for birdsong, but there is still so much music in the air. The mockingbird refuses to burst into song when the weather is this cold, but he lets out a deep *shoomph* as I walk by—whether a gruff "Hey, you!" or a quick warning for me to watch out, I'll never know. Trees full of starlings alight into miasmic murmurations that swoop through the sky against the red-gold trees. The final crickets of the year try their last fall singing before the first frost will kill them, unless they sneak into our houses to serenade us and eke out a few more weeks of life. Wind gusts tornado up the dead leaves into a percussive swish, like an exotic Zildjian still not designed in the family of cymbals. Nothing rare here, but it can all be beautiful if one wants it to and truly follows each emerging sonic form.

Those were some inhuman sounds I have deigned to notice. The noises of our species can also transfix us, for technology has always been best if it, too, admits the possibility of beauty. The whoom of a braking midnight train vibrates the Earth like the crashing ninepins of Rip Van Winkle's dream, that thundering trance that happened just up the river a few miles from here. Great oil barges humming up the Hudson, on their own or pushed by powerful tugboats. The waves these vessels make can surprise you, miniature tsunamis on the beach that might wash all your clothes into the drink while you're taking a swim.

Next, riding the train into the city, a busy mix of duct systems and rumbling of wheel on rail, hydraulic doors and slow screeching into the station, not smooth and efficient like a European high-speed train but a creaking old American thing. Reassuring in its irregularity, its inexactness sounds alive, like some strange beast, full of sounds we can't quite place so we can appreciate them all the more.

How much must we *know* about a sound to really get it? Do we need to forget the names of the things we hear to understand them as sounds and nothing more? Think of the word *acousmatic*, introduced by that old mystic mathematician Pythagoras to mean a sound whose origin could not be explained. These are the sounds that most grab our attention, whether or not we admit them as music. They are audible mysteries, and we don't forget them.

I had been thinking about this power of sound as the undulating unknown when I came across Brian Kane's fine book *Sound Unseen*, which traces the history of the acousmatic idea from Pythagoras on to phenomenology's quest to listen to the thing-in-itself before trying to explain it. What a beautiful contrast to our desire to know everything, to put a name on stuff to silence our internal wonderings. Kane's book begins with the story of the Moodus Noises, strange sounds that have been heard for years in the town of Moodus, Connecticut.[7] Being a Yale professor, Kane knew the story well—even the Indians talked about such things, calling them the voice of the god Hobomoko, who the Puritans later decided was Satan. Today's experts try to explain away the sounds as some form of subterranean seismic activity. The fact is, these strange sounds are heard every decade or so, and we still don't really know what they are. My point in telling you this is that the less we know, the more interesting they sound. What does this tell of us about certainty and soundscape?

The presence of the acousmatic touches us whenever we try to take in any soundscape we cannot explain. It happens all the time. Last night I awoke in the middle of the night and heard a whole composition in the dark from within my city apartment. Grounding everything was a strange warm beat. Was someone composing techno somewhere in the building? I doubt it. Was it the heating system practicing its own strangely clear rhythm? Along with it my wife was next to me breathing slowly and intently. A wash of white-noise tones came in from the window. Were they air conditioners in other apartments? They always seem to be on. Wind through the falling November leaves? A car driving through the rain? In the dark, one cannot know. Thinking about the acousmatic, I try to hear it first and foremost as a holistic composition. I remember Eduard Hanslick, bad boy of nineteenth-century music philosophy, who tried to convince us that music is only about musical questions and musical rules and shouldn't be heard as representing emotion or be based on any sounds of actual objects in the real world. Most of us are not such abstract listeners; we use sound to situate ourselves in our places, while sight floods us with information.

We usually want sound to refer to something real around us, unless we enter that acousmatic activity called "listening to music." Decide what you listen to is music, and part of you will stop trying to figure out what each sound refers to and instead tune in to the overall interplay of the sounds involved. If they are played by musicians or composed out of machines, the sense of disbelief is easy. If it is the sound of an environment or a normal walk through life, then it is closer to a form of life heightened by meditative insight. Before music, sound is sound. After music, sound is still sound, but our feet are walking a few inches above the ground. Paraphrasing an old Zen story, our attention to what resounds around us makes life just a little richer, keeping us that much more alive.

Hear the French sound pioneer Pierre Schaeffer invent musique concrète. He sits on an auditorium stage with a phonograph, manipulating recordings of a train, slowing them down, speeding them up, switching from one record to the next. Is it or is it not a train? Is it a rhythm, a noise? Is it beautiful, a curiosity? Is it created for us by accident or design? These questions ask music to be music, and I'm okay with that. The very word *acousmatic* names what should not be named, the unknown that should still surprise us, as technical-sounding as the sharawaji is evocative. Let's applaud the music of the world, the musicality inherent in all sounds once we decide they are worth attending to.

It's not hard to improve your listening. Just step back, breathe, leave space, turn off your phone, look up, close your eyes, and take in the sounds. They will be interesting. You will want to inhabit each one. Feel the music that can be found there and outline the forms. A field recording, no matter how fabulous, is only a marker of a live experience from the past. Music must be ever remade, and even what is made is never as great as what can be suggested. The notes that came out are only one possible subset of all the notes that could be played. A nightingale's song is one possibility of all the songs that could be sung. One city is full of them, and it is to there we must return.

# 7

## BERLIN LONGS FOR BERLIN

It's always a good idea to rewatch the classic Wim Wenders film *Wings of Desire* upon returning to the city of Berlin. It's one of few films that becomes more beautiful and different with every viewing. Its German title means *The Sky over Berlin*, and I remember how hard this vision hit back in the 1980s when it was released, a time when the Cold War and the Berlin Wall loomed in our collective consciousness, slicing this great city of excess and darkness in two. Empty lots. Rundown circuses. Gray graffiti. The importance of internal monologues that no one but the angels can hear. Peter Falk as Columbo, as himself. "I can't see you but I know you're there." The existential musings of Peter Handke's screenplay. *Als das Kind Kind war*. When the child was a child.

When I first read his novel *On a Dark Night I Left My Silent House*, I knew that one day I should release a record with that same title, and that I did, my first to be released on ECM. Berlin— yes, this gray city where I now have the pleasure of spending a full year of living easily, hopefully not emptily, with time on my side. Time turned into space. Memories of this film reverberate through me over the decades.

This was the defining film of our Europe-leaning college years. It describes so perfectly the yearning for love, that perfect woman on the trapeze, Solveig Dommartin. Back then we knew absolutely nothing about her, except that she was the kind of woman to compel an angel to willingly fall to Earth. Of course, for all of us lonely young men love was an unattainable hope, a longing for something we could not understand. It was like the angel's gaze in that film: once you saw her, your whole world would enhance as if suddenly going into color after thousands of years of monochrome. "Remember when we started coming to this place?" says Bruno Gans to the other angel, Cassiel. "There were swamps. Animals." In this black-and-white Cold War city only rabbits appear. Angels as the ultimate bearers of existential doubt, doomed to share everyone's suffering, barely able to alleviate it. There could have been nightingales!

Today it is easy to look up Solveig Dommartin and learn the sad news that she died of a heart attack in her mid-forties, younger than I am now. We were nearly the same age when she made the film. Wenders is a few decades older. He fell in love with her on the set, and they stayed together for the better part of a decade, she helping him on his epic journeying film *Until the End of the World*. After they split up she never worked in movies again.

The only time the film's internal monologues ever come to light is when Solveig meets the angel she saw in her dreams at the Charlottenburg bar after a Nick Cave concert. As she talks, he sits, transfixed, and listens. By the time she is done the whole place has cleared out save for the two of them. They are in love. Bruno has tasted the joy of being human. Berlin is no longer gray. Only the angels see it that way—on Earth it is alive in so many hues.

Every time I see this film I have forgotten the plot. There is still suspense. Was Columbo really an angel who fell to Earth thirty years ago—"Compañero!"—or is he some kind of devil trying to trick angels into signing on to a finite life down here, doomed to be full of tragedies and disappointments? "I can't see you, but I know you're there . . . I wish you could taste this coffee, smell this cigarette, feel the colors, lose your wings."

All this beauty couldn't save her. Solveig died too young, her greatest triumph having been on the trapeze in her perfect youth. We all die too young. My own parents got sick and suddenly their old age was changed; they did not get to enjoy gracefully getting older and the slow fading of ways. It was pain and suffering for ten years. They did not deserve that. No one does.

Now I seek refuge in sounds, a pure art of noises where nothing

FIG. 10. Nachtigall Imbiss (the *döner* isn't bad)

needs to mean, only be. Poetry is said to do that, but for me only music fits the bill.

The biggest change in watching *Wings of Desire*, now that I'm as old as those long-coated angels, is to be reminded of the power of love. Those shadows in the library! Their stark expressions and tiny ponytails behind gray hair, that short-lived hipster look of the 1980s. That part is easy to laugh at. But *love*—what kind of power does it really have? Do we listen to the angel or the devil? Can we even tell the difference? Experiences do not coalesce into time, they do not fully exist in memory. Time cannot become space, no matter how much reverb you wrap around the sound.

My search for ideas echoes in my head, swirls like particles of an unlocatable sound. The music must be in the structure, the string of thoughts, the explanation of tones. That's why I am so impressed by the nightingales of Berlin, how they sing tirelessly through the night above all the cars, trains, trams, parties, human laughter, and trembling bass. What sounds different for them calling all above the noise? What does it mean that so many of them are here, in this city of many cultures? They've given their name to at least one Turkish café. In between nightingales and kebabs I remember certain moments from my year in Berlin, those that embody the city through its sounds.

The sound artist Taras Mashtalir said an interesting thing at a symposium at the Wye on Skalitzer Straße: "Sound art is made to *sensitize* people to the sounds in their surrounding world." It exists to transform our perception, perhaps more so than other kinds of art. True, I have always felt that the highest compliment to be paid to any art form, be it music or particularly literature, is that it heightens our perception, lifts us on a cloud of greater knowing once we're done with it, as *Wings of Desire* does to me every time I see it. It is an especially literary film. It is all words, streaming inside the heads of the characters who think them, heard only by the angels and the cinematic audience. Otherwise, the words on the screen travel in silence.

But sound art, Taras seems to be saying, is far away from music. It is a perceptual experiment that wants us to hear the world in a new way once we leave its presence. Instrument builder Ken Butler described an imaginary example: "If I go into a gallery and there is a thrumming tone that can only be heard when you are inside an invisible figure-8 space in the gallery—technology can do this and I have heard it—that is a valid, amazing artwork but it is most definitely *not* music. It is not pretending to be." It is an aesthetic experience realized in sound, all the purer the less anything visual surrounds the sound. It is not to be recorded, not to photographed or otherwise described. It is sound in a space that you must inhabit in order to believe.

That longing in Wenders's film, the unattainable beauty of the trapeze girl, is changed into the constancy of warm days. I strive for those days and long for the hours in a bright red bed in Berlin, perfect for warmth, while the gray days that are not quite days loom on outside the window. There is no essential pull to go out. The days reverberate within. The hunt for a story that needs to be told. A story deeper and realer than any I have previously offered, with less information. The true power of the need for sound.

One day Markus Reuter came to visit. He is a soft-spoken, ex-tremely thoughtful musician, an instrument inventor, composer, and producer. A master of the touch guitar, an instrument played with the fingers of both hands, no strumming, just tapping dually on the fretboard, Reuter began learning his craft with masters such as Robert Fripp, and today, twenty years later, he has joined his teachers in the latest incarnation of their virtuosic instrumen-tal rock music, the Stick Men, an offshoot of the legendary band King Crimson. (We can age to become our idols. That's one fine thing about music.)

I have always wanted to play with Reuter, and he has finally agreed to come to my studio in Berlin. Sadly, he arrives with no guitar in hand, just a mysterious shiny vinyl backpack and match-ing shoulder bag.

"What gives?" I ask with disappointment.

"*Ja*," he muses, "I thought about bringing a guitar. But I thought of all the music I have played this year and thought to myself, I have said all I can for 2013. I won't pick up the instrument again until the New Year. Then we shall see. Instead I will take up your sound, and transform it. I make so much electronic music but I absolutely *hate* synthesizers. For me the sound must begin with something acoustic, something real, an impulse from somewhere. Then I will capture it, and transform."

"Yes, I know what you mean," I say. "I, too, have always wanted sound to *live*, to be alive. I want my machines to cry out like ani-mals. We should fear them, empathize with them. I don't know if they are fooling us or admitting our or their true natures."

"What does that mean?"

"I don't know. Sometimes I need the unevenness of a sample from reality to begin. But samples have blurred now into artifice, and the electronic. Sometimes purely synthesized sounds, treated unevenly, begin to sound alive. But I am still trapped in the para-dox of music on the machine. No program on it has ever truly sat-isfied me. I yearn for the future. I do not want to emulate. I want a sound I have never heard before. I should not be able to hear what is happening. When it comes to effects, I want the ultimate

... something that cannot quite be explained or identified. No re-verb, chorus, distortion, echo, or clear loop. The amazement of the beyond, the perfect realm of pure sound. I don't know if it exists. I don't think I'll ever find it. Longing keeps me going, there is no satisfaction. The more futuristic we want to be the more we date ourselves and create something that will age." I worry that tech-nology will sound old too fast.

Markus doesn't buy it. "What are you talking about?"

"Sometimes I just don't want to create any more unclassifiable sound. Music has become too easy. Though I guess it's only easy if you can articulate what you like, or at least manufacture it."

"I guess I agree. I suppose that's why I decided to stop playing for a while. I practice, but I will not play."

"What's the difference?"

"Practice for me is going for greater attention. I do not practice scales, patterns, exercises, *chops*. I practice only the moment of what is barely possible. The leap, the fluidity. The river of the tone. I listen further to hear what is barely there. The beginning and the ending matter most. Who really pays attention to the middle?" He laughs. "You can learn so much about a musician by how he starts and ends his notes. There must be commitment in the jour-ney from the beginning to the end."

"Can you always tell when the commitment is there?" I know I can't.

"I think so. But it is dangerous. Musicians are always making deals with the dark side. They are trying to say something that cannot really be said, not only not with words but not really with music either. They want to sing the darkest hollows of human vir-tuosity. Music at its highest level always has the power to consume you. We all get lost in it, and then hardly want to speak to one an-other. When I am on tour with Pat Mastelotto and Tony Levin. we might ride a tour bus for seven hours across the Great Plains and hardly speak to one another. It is in those long silent rides that I realize we are all three of a kind, that we truly do get along."

No reason a musician should be expected to communicate in any way but through the music. Why do we even pretend to talk? Why do I keep writing books about something that communicates

PLATE 1. Thrush nightingale singing in Helsinki, Tuulisaaren Park

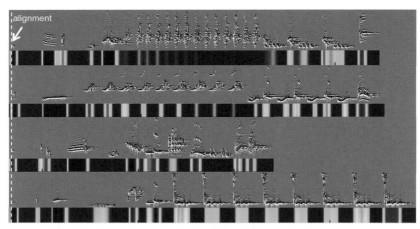

PLATE 2. Visualizing the continuous amplitude of aligned songs, by Tina Roeske

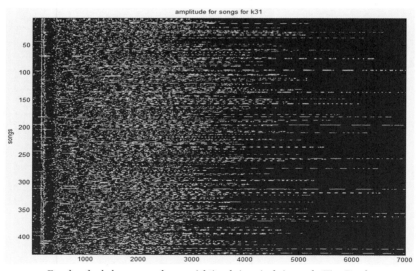

PLATE 3. Four hundred phrases sung by one nightingale in a single image, by Tina Roeske

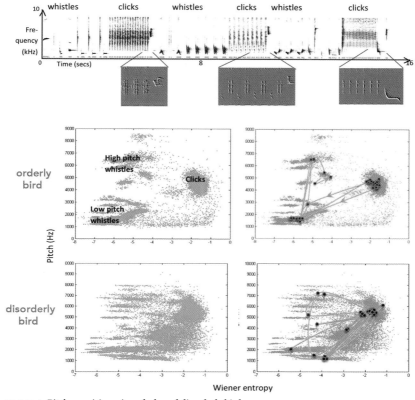

PLATE 4. Pitch vs noisiness in orderly and disorderly birds

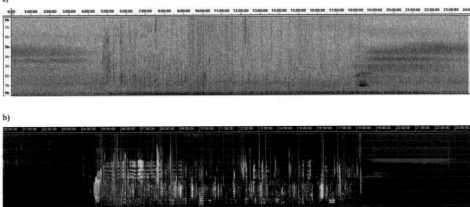

PLATE 5. 24 hours of an entire soundscape categorized, by Michael Towsey

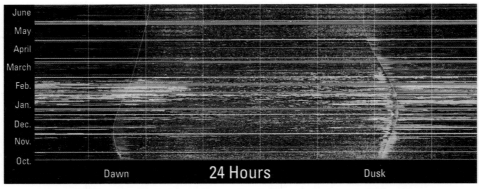

PLATE 6. Eight months of a single soundscape categorized in a single picture, by Michael Towsey

PLATE 7. Bernie Krause's "Great Animal Orchestra" installation in Paris

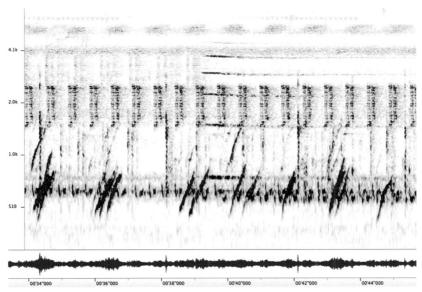

PLATE 8. The most beautiful sound in the world? A Borneo soundscape

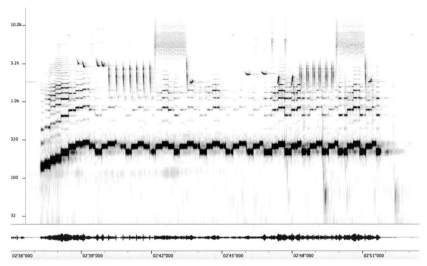

PLATE 9. "Sharawaji Blues," early morning in Helsinki, clarinet and nightingale

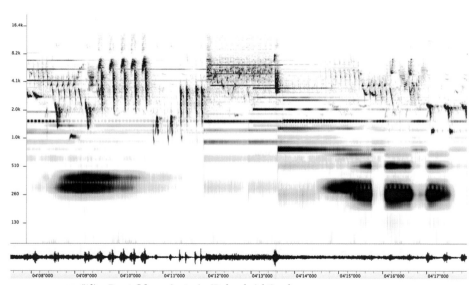

PLATE 10. "Alien Beauty," four minutes in, iPad and nightingale

PLATE 11. Blyth's reed warbler

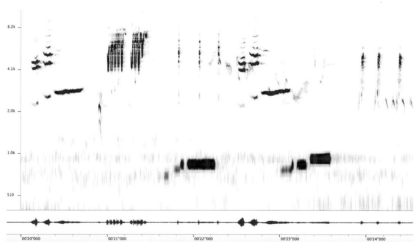

PLATE 12. The Blyth's reed warbler teaches me a tune

PLATE 13. Nightingale singing in Volkspark Hasenheide

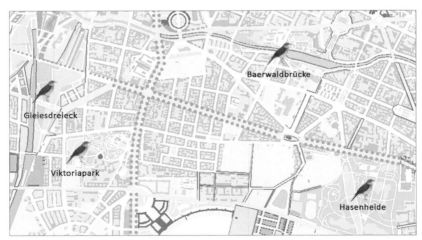

PLATE 14. Where we found the birds

so much more poignantly through sound? Reuter spent years practicing, honing his craft, helping others improve theirs through
his keen ear and production skills. He picks up on things in the
sound that no one else hears. He works methodically, his fingers
at the laptop, tweaking the details. "Look at all those huge towers
of effects Robert Fripp lugs around the world," he smiles. "Who
needs it? … It's all for show. Tools far more powerful than those lie
right here in this little machine."

He's right, and therein lies the problem. We have so many precise musical controls at our fingertips and the ability to produce
such amazing sound so fluidly that we hardly know what to do with
this ability. We should hone our listening skills and our sense of
aural connoisseurship to push music forward. Music is changing,
as are our tools. Yet our desire to be delighted by the rich beauty of
sound remains. The surge of all possible sonic choices encourages
ennui, but all the same we rarely know what it is that we hear.

I celebrate the possible love of all noise, as also the critique of all
noise. We learn to like different kinds of sounds and always wonder how they might be *better*. Legendary enemies of popular music
such as the philosopher Theodor Adorno say it's all just the same-
old same-old masquerading as the new to give us all a little pleasure without ever really rocking the boat of what music is allowed
to be. Those who love it say otherwise. You are the music while the
music lasts, and it does last. The echo resounds. At its best it will
be all that has ever mattered.

On December 7, 2013, an historic concert took place at the Akademie der Künste. In celebration of fifty years of German DAAD
artistic grants, the three original members—Richard Teitelbaum,
Alvin Curran, and Frederic Rzewski—of the ensemble Musica
Elettronica Viva came together to perform again. Two of them I
had met numerous times before: Richard because he lives near
me in the hills north of New York, and Frederic because he was my
teacher once at the Banff Centre for the Arts in Canada more than
thirty years ago. These guys are all in their mid-seventies, with dis-
tinguished avant-garde careers. Curran taught at Mills College in

California for years, all the while managing to live in Rome. That's quite a commute.

Each of these three lions of the avant-garde has spent a lot of time plying different experimental trades, and all, like me, used teaching at universities as their main source of income. Teitelbaum has always been one of the subtlest players of the synthesizer I have ever heard, favoring quiet textures and tones over loud experimental outbursts. Rzewski is a political firebrand, his most famous piece an hour-long set of variations on the great Chilean revolutionary tune "The People United Will Never Be Defeated." Curran plays a keyboard full of pithy language samples and occasionally fragments of show tunes on the piano. All three look like hipsters who have, improbably, survived and keep doing what they do.

"I don't know whether what we played was good or bad. I suspect a little bit of both," said Rzewski, repeating a line he'd said over and over again. I last saw him twenty years ago and he looks as though he barely remembers me. As though he is only pretending to know who I am. Isn't that how it always is, when you go up after a show to speak with a musician you admire, someone you don't really know but you have had a handful of encounters with over the years? How much do these encounters matter?

Rzewski would probably agree with me. He has always been a committed socialist, at least in the politics of his art. He smiles after a lifetime of belief: "Socialism was not a bad idea, you know," he sighs. "I mean ... Russia used to have the greatest tea in the world. After the fall there were billboards everywhere saying 'Lipton's.' Can you imagine? They were now all supposed to drink Lipton's crappy tea. The revolution was over. Capitalism had arrived."

"We thought we were so unhappy," laughed a Hungarian composer sitting at the table, who'd been around the block and then some. "Turns out we didn't realize how happy we were before." I'm hearing these stories in the café of the Akademie der Künste, an erstwhile and austere temple of culture at the edge of the Tiergarten in what was formerly West Berlin.

Most of the patrons here are established and elegant, well-dressed in an understated but slightly shaggy Berliner way. All the women have red or green glasses to accent their smooth blacks and grays. There is a consistency to the art they are celebrating. Rumbling sounds and super-bright lights from yellow to blue are assaulting us in one piece. We have to wear what we thought were 3D glasses but actually are cloud glasses, just white translucent paper that makes everything muted and blurry. Think of how much money James Turrell could have saved if he just made all visitors to his exhibitions put these on while he shone bright spotlights into their faces. It is something to think about: an easier way to get the pure experience of light.

Curran was playing fragments of somewhat familiar songs and rhythms but would never hold on to any one of them too long. A groove was an impossibility. That would perhaps be *too* populist, too mainstream, a direction electronic music surprisingly took after this trio's noble and ancient experiments. Repetition has kept us moving to the beat for thousands of years, as we learned from insects and ocean waves. In nature, cycles may be predictable but are never precisely the same. Machines make repetitions inviolable. "When there are loops," said one techno producer in a Berlin magazine recently, "all is well. In life, in relationships, in music." Habit, John Dewey would have said.

At the music studio at Harvard in the 1980s, our loops were actual bands of tape hanging on a wall. We would pull them down and string them up on specially modified tape recorders woven all across the room on empty microphone stands. If someone else in the class was stealing loops, it meant physically pulling a loop off a peg and using it for one's own purposes. Now it's all virtual, immediate, cut-and-pasteable right on your phone.

Musica Elettronica Viva still use no loops. Either they are purists or they don't want to offer themselves up to the aesthetics of the beat, a world of genres and subgenres that sound identical to the uninitiated: house, jungle, trip-hop, techno, glitch, dubstep. Such names need to be restless, always on the move, like the songs of humpback whales that change from year to year, month to month, week to week. You don't want to be adopted by any of them if you want to be a "new" musician, ever afraid of a club that would have someone like you for a member.

I can think of a few ways their show could have been *better*. First, divide the improvisations into distinct pieces, so as not to feign its being one giant coherent statement. They called the thing "Symphony 104," in honor of the 104 years that the American composer Elliott Carter lived, composing difficult but sublime works to the very end. Indeed, many experts felt that he composed his best work *after* he turned one hundred years old.

Secondly, I would make sure the electronic instruments never got the better of the acoustic ones. Or I would have had all three playing complex overlapping murmurings of sound while some acoustic instruments overlaid that wash of sonic possibility, all so quiet that you could barely tell which of them was playing what. I would keep each of their sounds spatially located, so that after a while you *could* tell who was playing what and you would start to hear the electronic personalities of each of the Viva musicians, hoping to hear true maturity, a sense of possibility that only fifty years of music making can suggest. I would like to hear in their music my future, finding hope that I can still be doing what I am doing now thirty years later and still have deepened to the point where I don't feel ridiculous for carrying on.

The colors of the 2014 Club Transmediale (CTM) Festival are the unmeldable opposites of red and green. Their brochures print red on green and green on red, colors that are barely readable upon each other. Wittgenstein wrote that such a color as "reddish green" (or "greenish red") doesn't exist, and that anyone who uses this combination is force-feeding us colors that don't blend, like sounds never meant to mix with each other. How much electronic music can one's ears really take in? Even before entering the concert space, our heads are ringing.

The future is not the resuscitation of the old sinusoidal "warmth" and the robot vocoder fuzziness of childhood. Sometimes I think it is the grand mashup of all sounds we know or should know, all our likes and dislikes piled on top of one another until we are pushed beyond our opinions to the edge of change. I know I like noisier structures after listening for hours to crickets late at night in a field. I smile when these songs of Earth end up so much like the potentialities of machines. There are easy secrets to make electronic music warmer and more inviting, and they are more than just a matter of tone quality. There are three easy tricks. Trick one: don't have a continuous thumping beat. Leave space! If you want to make a song out of a repeating phrase—this is something machines *love* to do, and we know it is best to keep them happy—then repeat something at first unfamiliar. Something you don't want to dance to but after a while you find you can't keep out of your head. It may take weeks or months to find such a perfect, uneven loop.

Trick two: the emotional, mournful sound. A single remote chord held over longer than you at first think you need. The tug of an ethereal minor chord never fails to move. This is why people warm (or do they cool?) to my piece "The Killer," with its mournful repeating calls of mammal-eating killer whales, all singing above a droning dark guitar.

Trick three: surprise. Just when you think your listener has figured it out, add something unexpected. Throw in a curveball, move the music or the story to a place that no one thought would come next. I know this is a contradiction to another view I often hold, that music is actually easy, that the simple repetition and the comforting full tone resulting from years of preparation and practice is enough to lure people in. But then you must take them somewhere they do not expect to go. Don't let them second-guess the music. It is even best if you have an audience that is hostile, that doesn't think they are going to like you, and then you convince them slowly, by invitation, respect, and then the twist.

I go out into the street to find a sound that encompasses all other sounds, walking from the Café Tagesbrot along Dieffenbachstraße to the Turkish Market on the Maybachufer. A strangely warm day, and the first pure blue sky in many months over a city known for its infernal pall. People are struck with silent astonishment at the extraordinary weather. Birds are waking up, flowers suddenly blooming. Spring seems to be coming four months early. That same day in New York City, the temperature has plunged 50 degrees Fahrenheit in four hours, a phenomenon that hasn't happened in ninety

years. I wonder what *that* sounds like? I walk down the street, trying to figure out if the beautiful weather actually *has* a sound, a key
tone of the day. And then I hear it, a barely audible music. Someone is playing a xylophone on a balcony many stories overhead.
Perhaps a child, perhaps a mobile phone's charming ringtone. And
I think to myself: is it really music I'm hearing, or the background
sound of the city itself?

I am ennobled by the shadows of giant wood pigeons sitting
silhouetted against a black metal cube atop the chimney of the
neighboring building. A ladder leads up to the cube, suggesting it's
a support railing for chimney sweeps as they inspect the channel.
But how often do they need to do that? Once a year? Maybe it is
designed for birds, to give them the highest point on the building
to survey their domain. These wood pigeons amaze the foreigner;
they are twice the size of our American pigeons.

I bring this image up not only because it is stark but also because of the sounds these giant pigeons make. A low *whoomp*
*whoomp whoomp*, not quite as remarkable as New Guinea's gigantic Victoria crowned pigeon that one sometimes hears in aviaries,
but a resounding boom nonetheless. Those low, visceral sounds are
missing on that sunny winter's day, replaced by the xylophone in
question. It's an arresting if imperfect sound, lacking the full range
of frequencies where each knows its true and correct place.


I often think about the sound of wind. It's difficult to record yet one
of the easiest sounds to remember. The howling whistles through
thick concrete walls in old Eastern European apartment blocks
are always more dramatic than the actual air. The heavy gusts on
the abandoned Berlin Tempelhof Airport just down the street are
always far quieter than they feel. I am never only interested in the
sound of wind, or of anything else for that matter. Nor am I seeking a politics of sound, though I know it helps sell the work: what
if freezing ice sounds better than melting ice? Shall we then put it
into a work that comments on the menace of climate change? It is
not enough to be thinking about ice. It is not enough to be chasing
ice. You have to do something beautiful with this concern.

FIG. 11. Lars Schmidt at the Tempelhof Airport

If these words seem cold, it's because they cannot contain the emotion I feel. Either that, or my feelings are cold and the words track a lifetime love of coolness, as articulated in the writings of everyone from W. G. Sebald and John Berger to Steve Erickson and Lao Tzu. What might make *you* want to turn the page in this rambling quest? The search. The search for the perfect sound, and the perfect attentiveness needed to find it.

Xylophone on the balcony, silent spring. Wind in the airport blowing me down, but silent, as there are no trees for it to bend. I nearly slip on the ice on the second runway. I cycle as fast as I can toward the freestyle garden encampments near the Flughafenstraße. There I once sat in a chair rimmed with clear plastic to protect myself from the wind on one of those bitterly cold autumn days with my friend, the tango instructor Lars Schmidt, eating *Bergkäse* on a slab of dark organic bread. "I have to eat out of doors," he explains as we leave a warm café for the howling sunset breeze, on the plains where the Luftwaffe used to rise and cabbages and rapeseed are now grown. Out of the wind I can finally hear the wind. Air only makes sound *against* something; the plastic tarps are rattling up a storm. The food tastes fresh, and we shiver as the sun goes down. Of course the sandwich tastes way better like this. Lars is right.

He'll be leaving Berlin soon. His warm clothes are stashed in a Volkswagen bus at his ex-girlfriend's parents' house outside of Marseilles. A long way to go to grab a sweater.

My nightingale city lies beyond the realm of its possible sounds. Next to the old runway a group of scroungers roasts a wild pig whole over an open fire, hiding behind a few garbage dumpsters on a balmy day in the middle of winter. The sky, as if dreamt by one of those cultures that have no name for blue, is no particular color at all. Indeed, you hardly sense it is there.

The next day I return to see what is left of the boar. No sign of feast or beast. But Tempelhof is always an interesting place to go. I wish I could tell you I went there for the sound, that the sound was tremendous, that there was really something to hear. There was not.

I noticed more the smells, via the one sense humans really do not know what to do with. The Bahlsen Cookie Factory is just south of the airport, and its warm, buttery scent wafts over the southern runway. Next to it is a giant old hangar with a sign on it that says "Tonfilm Studios." Whether current or an ancient relic of the first Deutsche Filmindustrie, I don't know. The airport is always sublime, a bright beacon of space in the otherwise all-settled city. What a piece of real estate, say the developers. What a park of the people, say the people. Before seeing it I envisioned something gigantic, all paved, hot and sweltering and raw—essentially what it would be like to wander around Kennedy Airport in New York if one day it suddenly shut down. Tempelhof is earthier, with a history. There are wild community gardens and dog parks sprouting on either side, more Nightvale than O'Hare. It's a strangely comforting place. Everyone smiles, whether rollerblading, flying kites or copter drones, or roasting pig. I smell the pig in my thoughts, imagine the sizzling I never heard ...

At an evening dedicated to artists on the Gruenrekorder label, which only releases artists making music out of the soundscape,

Peter Kutin put us all into darkness, and then the sounds began: ominous, rumbling sounds, strange sounds, the results of simple contact mikes stuck into sand and onto old railroad tracks. These are very inexpensive microphones made by soldering wires to simple piezoelectric discs, low tech and famous for producing strange, abrasive timbres.

Then, suddenly, he flashes images of the places he recorded. Stark, nearly monochromatic images of Chilean deserts, empty beaches, abandoned greenhouses, and deserted salt mines. The choreography of the images and the sound is perfect, even if at moments I must stick my fingers in my ears. It's a *piece*, with a beginning, middle, and end. In a sense I count myself at fault for tainting the whole evening's event by urging all participants to speak about their works. I'd worried the whole thing might otherwise come across as long blocks of decontextualized sound. Although my worries were unfounded, those long blocks might have sounded better than what we actually get. Some participants speak too much, as if imagining their works were more interesting than in fact they are. Not all artists should talk about their creative process. The better works speak for themselves: in this case, Peter Kutin's. The others have been steeped in too much explanation. Sometimes we musicians must learn to be quiet about what we're doing, or at least to know when and how the work should be allowed to speak for itself.

Kutin tells the story of an old mountaineer who fell into a crevasse and was stuck in absolute darkness for six days before he was rescued. He survived by catching trickles of water dripping from above into the abyss, and by doing his best not to fall asleep. Since he was in complete darkness, it was the perfect subject for a radio piece, says Kutin. Next, he's working with the Austrian director Daniel Hoesl on a film about a future Earth on which humanity is no more. What will it sound like? They are spending several years filming the ample ruins of human civilization in places of complete silence. There will be absolutely no location sound in the final film. Everything will be created from scratch, like a homemade cake.

I'm waiting in the front of the line at Stadtbad Wedding, a series of big old swimming pools turned into a music club for the CTM Festival. "Hold on, please," glares the bouncer. "Please do not step on the landing, back behind the line," he instructs quietly. I do as I'm told. Is the place already packed? In ten minutes we get in through the first door, then go through the full-body patdown. No weapons or recorders among us. In fact, hardly anyone is inside. It is still early, only 11:00 P.M. We are happy to find that Chris Salter's Xenakis installation is still up, in one of the empty pools. A swimming pool is a disarming surface to navigate. It is slanted and dry, not slippery, except that lots of small pillows are poised on this one. There are long, straight radial wires with bright noble-gas lights pulsing on and off. I believe it is an old technology, the same kind of thing Xenakis and Le Corbusier used in the famous Philips Pavilion at the Brussels World's Fair in 1958. The sound track seems sampled from the sounds of the urban world, moving around our heads and revolving in and out of focus. You lie down on the slope and feel the geometry of its microcosm. Occasionally a razor-sharp laser beams its pure green line through the space above your head. Here is an oasis of sonic peace. And yet with the dotted lines drawn through the room, I am reminded of Xenakis's mathematical drawings, doodles of precise obscurity that only make sense today in a world he didn't live long enough to experience. I remember that he designed a strange concrete summer house for François-Bernard Mâche, a French composer who holds the Messiaen chair in Paris, the one heir to Olivier's obsession with birds, a man who wrote a book called *Music, Myth and Nature, or The Dolphins of Arion*, one of the pioneering books on the subject at hand.

Outside the swimming pool the crowds are building. A mysterious sign is duplicated all over the halls: "Due to illness Boddika will be replaced by Shackleton." The stairs go up and down through corridors illuminated with the glaring reds and greens that are the official colors of this festival, and these tunnel-like hallways that used to be locker rooms fan out into impromptu alcoves where various bars have been set up, all ready for the big crowds that are no doubt on their way. In the boiler room, each DJ seems to have his own table full of gadgetry, all covered by grungy painter's cloths, waiting for their turn when the cloak will be removed and the magic unveiled. In the biggest swimming pool appears Jeri Jeri, a band of Mbalax drummers from Senegal. Their record, as realized by producer Mark Ernestus, is a cool hybrid of the electronic regularity of Berlin electronica with the gut reality of a stage full of actual drums, but the live show is the pure West African sound. How can a machine at the bottom of a pool compete with such raw power of? Grooving along to this band at 2:00 A.M. makes me realize why it was worth staying up so late and not noticing the time at all. And I wonder why we bothered to invent all these machines when the actual groove is so much more alive with real people, deep in the beat but never exactly the same. Why did I ever start playing with that Arp 2600 in the Wesleyan Music Lab when I was seventeen? My clarinet through its ring modulator sounded awfully strange. Why would I think such a distortion would be in the least bit better than the pure sound of my lips on the reed? I remember that one time I played with Scanner at Tonic and the clarinet was so badly miked it sounded distorted the whole show and several people came up and said how much they liked it that way, that they had never heard a clarinet sound so gutsy and hip. This led me to think that distortion proves that listeners like roughness more than clarity, even though the incoming sound is degraded and pushed a bit further toward low-fi, if not even low-bit, noise. But again, we like noise. We prefer it to the pure sounds of which we know our machines are capable.

Why is that? Perhaps because we are ourselves impure. Only our ideas beget pure circles, squares, and triangles, while the real world offers the unevenness of fuzz. Grit, glitch, grunge, grime. No actual surface can ever be purely clean. No real human voice sounds exactly like another. That is why we all must sing.

And why we all must listen.

Rhythm is an easy secret: just leave enough space for something uneven to repeat, and in our need for even unevenness we will groove to it until we figure out the trick and grow bored. Is music easy? Not really; but music that is too hard will only be appreciated by those who take the time to know. How much will you pander? Depends on whom you want to reach. I don't know enough about pop music to care either way, or maybe I'm just too old to care, to be so swept up with music at the birth of passion that nothing else seems to matter. I love describing artworks that are too great to describe. Like that all-white painting in the Yasmina Reza play *Art* that is too great for anyone to actually see, or the glowing white square in Lisa Cholodenko's film *High Art*, we must imagine it. So many examples like this cry out to me, like the boy in Joseph Vilsmaier's film *Brother of Sleep* who is asked at a competition to improvise upon Bach's chorale "Schlafes Bruder" but has no idea what that is so he just plays his heart out. It's 1799 and he's just come down from the squalid hills to play a proper organ for the very first time, and wouldn't you know it, the roof of the cathedral is blown clear off by his sound! Or is it only a metaphor for untutored greatness? Like the fictional Vinteuil Quartet, in Proust, composed by Hans Werner Henze in Volcker Schlöndorff's film *Swann's Way*, a scene I remember best because the society matrons in the audience seem to jerk and spasm in some kind of restrained autoerotic dance brought on by this most avant-garde music. The thing is, Proust is trying to describe these experiments as if they were grand emotional moments, passionate sounds that genuinely touch us and are not mere mathematical experiments made by composers who, as David Stubbs tried to argue in *Fear of Music*, insisted specifically on making music that no one was supposed to like. They were to be, as Schoenberg famously said, "the elect who refuse their mission."

Do you agree with that? Were we all taught in school to make music that would scare people off?

An incredibly loud subsonic sound rumbles from the giant speakers and shakes the entire room. It's not any kind of beat but a huge trembling drone, honestly painful to hear. I jump at its intensity. A woman next to me turns and says, "So, do you fear the music?"

I fear the noise until I am able to hear it as music. A sound so gigantic as to seem unreal.

I write to Mark Ernestus because I am intrigued about exactly what he has been doing with these Senegalese drummers. The live show sounded fairly traditional, while the recording has various subtle delay effects and a remixed quality.

It takes more than a month for him to respond. "Come to my studio," he says cryptically. "I will show you. Just don't tell anyone where this place is."

Ernestus tells me he has gotten different kinds of musicians to play together who would otherwise never mix. Yes, it is traditional, but it is also a blend of unlike traditions. Sometimes there are conflicts. Some of the drums remain in this studio because one member of the band was cursed—he did something very bad that no one dares name. No one touches his drum now; it sits unplayed by the wall. What would happen if someone played it? "We don't know," he muses. "Some kind of bad craziness."

There is a whole wall of modular synthesizers covered in vinyl slipcovers. There are giant speakers that look like something out of the film *Metropolis*. Neither of these look as though they have been used in many years. On the other side of the room is a more typical computer with small home studio–type monitor speakers.

"Why don't you use those giant, imposing speakers over there?" I wonder aloud.

"Very simple," says Ernestus, founder of the Hard Wax Record Shop and a pioneer of the Berlin techno scene. "Those speakers do not represent reality. Let me show you." He flicks a few switches and the mix is rerouted to the ancient ones. The sound is incredibly balanced, a firm, steady bass, a midrange that touches your heart, and crisp highs. Impressive. "You see, what is the point? The real world does sound like this. A recording mixed on such a system wouldn't stand a chance in the actual world. That's why I switch over to these more normal monitors here. There is no reason to deny reality."

Nothing should be more real than the music. The system cannot be greater than the system. What is rare can be listened to by all. We just have to know how and where to find it, and what to call it.

FIG. 12. The fallen dinosaurs of the Spreepark

In Berlin I long for Berlin. The blackbirds began singing four hours before dawn, and skylarks hover virtuosically above the Tempelhof Field. The camels have awakened after a dark winter in the Hasen-heide, and they're chewing up old Christmas trees like all the other animals. It is only a matter of time before this year's nightingales arrive and my deep contact with this ultimately twittering bird will be tested once again.

I seek a delicate beauty of tone, on the clarinet and in the stream of words, but I do not know if I will really get there. This is still not the text, just a precursor to the text. I will need to pile on images and accidents before letting the true story go.

Did you know there is a boat that takes you and your bicycle all the way to the Müggelsee? Can you believe that the Ferris wheel in the long-abandoned Spreepark Plänterwald sometimes turns in the wind? Next to it the dinosaurs lie down to listen.

Maybe the whole place will soon come back to life. Lars Schmidt is right: even in the city, food and music taste so much better out-side, in winter, out at the former airport, where the wind whips through home gardens' late-afternoon light and sound and sense cannot be believed. In the city we try so hard to evoke nature through our words, our desires, our desperation, our longing to get out in the countryside. But we're still here, waiting through those long winter months for the nightingales to return. In want of sound, we dream of sound. Close your eyes, put on the headphones, listen to the world of possible clamor that can be brought in from the field ...

# 8

## ELEVEN PATHS TO ANIMAL MUSIC

I was about six years old when we moved from New York City to the countryside, and for years I didn't have many friends but would take long walks by myself, wandering in the woods and by the river, listening to all the bird, bug, and water sounds and wondering how I could ever fit in with the natural world. As a teenager I learned that nearby were some musicians who had formed a commune in the forest to make music with birds and wolves. That group was called the Winter Consort, and I met their leader, jazz saxophonist Paul Winter. I told him I wanted to join his group and he said, "No, we are not the people for you. You will have to find your own path into the sounds of nature." It took many years but eventually I found my own way into playing live with these more-than-human musicians, starting with birds, then expanding to whales and even insects.

When I was seventeen I had a summer job tracking birds in the Sierra Nevada mountains of California with Dave DeSante of the Point Reyes Bird Observatory on an Earthwatch trip. Each morning I had to go out on a particular subalpine hillside and take notice of all the birds, some of which had little bands on their legs so we could identify them, and write down their detailed movements on a map. Why we were doing this I can't quite remember. Usually they were small finches and warblers, but one day I was tracking the movement of a large goshawk, the kind that liked to eat these tasty little birds. Tracking a big, exciting bird like this was a lot more fun. As he flew all over the place, I drew big swaths across the map.

At one point I lost him, so I sat down to figure out where he must be. I heard a rustle in the leaves. He was sitting on a branch right above me, looking down at the map where I'd been tracking his movements, as if he'd figured out what I was doing, much to his displeasure.

I laughed, putting down the map, and picked up a small pennywhistle. I listened to all the birds singing around me, and it felt so great to be out there doing this strange job that suddenly it all made sense. I started to play along.

That was probably the first time I made music with birds. And the last time for many years, until I took a walk in the 1990s with Canadian composer R. Murray Schafer. We were circumambulating a lake in Banff National Park, and he was explaining how he had composed a piece for trombone, soprano, and wilderness lake. The singer was to float across the lake in a canoe while the trombonist was hidden in the forest. The piece took place at dawn, and the audience members were scattered all along the shore. The piece was written to leave space for all the natural sounds of the birds singing as the sun came up. I thought: this is a piece of music that actually makes sense to me, something that would set the stage for human beings to culturally find our place in the natural world.

I am an improviser, having learned from jazz, Indian and Tibetan music, and all kinds of sounds that have crossed my path. I knew that the best thing about jazz is that it is open to other traditions, that it makes room for sounds and structures it has not encountered before. I knew I wanted to develop the details of how to improvise together with musicians of species other than my own. Birds were a natural choice, and nightingales the perfect ones to work with.

It was in Helsinki in 1998 that I heard the nightingale for the very first time, because they do not live in North America. I was teaching a course on the philosophy of sound at the University of Art and Design, now Aalto University, in the Media Lab. Walking home to Kallio in the middle of the night in late May, when it was barely dark, I heard this incredible birdsong. It was louder, more energetic, and more complex than I believed was possible. One day, I thought, I will come back to Finland and make music with this bird. Eighteen years later, I finally did.

I sit connected with the world of information like everyone else playing games of avoidance. I am ordering books on nightingales in various languages I can almost comprehend, knowing that I have so many books on the subject on my shelves I can hardly keep track of them. Who needs everyone else's nightingale stories when I have more than enough of my own? The world-music sage Ben Mandelson might say, "We always get the nightingales we deserve." So if I am meant to find an elusive one, then that I will do. If I need a feisty one, then such a bird will appear. If I am a tawny owl, I hope to swoop down upon one from above, gliding in for the kill.

In June of 2016 I returned to Helsinki to film human/nightingale music with an ear to several advantages. The city is very far north, and the birds arrive a month later than they do in Berlin. Helsinki is at a high enough latitude that most of the night is light, with no darkness to impede our ability to see and get the music on video. This makes the birds quite agitated, as it's harder for them to hide. In Finland I never saw them sit still; they were always darting around in the bushes like Muhammad Ali in the ring. Flit like a nightingale, sing up a spree. Here I was taken back to the moment eighteen years earlier when, one bright midnight, I first heard this amazing bird.

My favorite recording from Helsinki is almost the last. After finding out that the best times to play are 10:00 P.M. to midnight, then 2:00 to 4:00 A.M., we tried to sleep two hours in between, then take a long nap from 5:00 A.M. until 1:00 P.M., something like jazz musician or club rats' hours, but definitely exhausting for regular folks after too many days of it. After six days I am totally fed up with the madness of it. This one bird has been hiding from us on the west shore of the island, with the distant waning traffic in the background as darkness finally comes near midnight. We can't seem to film this character, since he darts around in the bushes, by now well aware of what we have in store for him. I am so worn out by the vast distance between my human music and his *luscinia* world that I almost give up, only fingering my keys and rings to make beats and rhythms as he does. It's cathartic, clearing my sonic palate of whatever expectations remain.

I'm so fed up with this weirdness that I start to repeat a looping chromatic melody, which soon moves toward the blues (plate 9). Does every species get the blues? I've always felt that secret wolf tone in between major and minor subverts those ridiculous categories of happy and sad that threaten to trivialize all human music. No song is actually happy or sad. All music is somehow in between, relentlessly inside emotion chock-full of *boori* sounds, impossible to divide up into mere laughter or tears.

The song of the nightingale remains uncanny. It is nature's electronic music, oscillators and tones and rhythms and noises that remain outside the rules of the Western canon yet that are still obviously musical, if beyond our ability to say precisely why.

The transcription of nightingale songs into easy melodies might be an equally bad idea. If the bird uses scales, they are not our scales. Its musical categories lie beyond the human sphere. Click, whistle, ratchet, *boori*. Wolf tone, hook, bend, swerve, slight, riff. Words never emerge to cover the coolness of music. This bird's song is relentless and uniquely strange. We will never get to the heart of it. I remain amazed how few books there are about it. One in Dutch, maybe two in German, three in English but all by the same writer, Richard Mabey. One poetry anthology, put together by a man who lives far from anywhere nightingales can be found.[1]

This is a song, like so much inscrutable music, about rhythm bouncing off silence as the source of the form. Does the bird sing hundreds of songs, or one long song composed of hundreds of similar yet somehow different phrases? Are their pieces, like concerts of Hindustani music, single giant compositions, worlds the listener and performer may enter, that honestly have no beginning and no end? By joining in, we may come to know it.

Cellist Beatrice Harrison smiled when the nightingale seemed to change his music in relation to her famous phrases, in her garden back in Kent in 1928. Yet she herself didn't think to change her music in reaction to the bird. There are human ways to learn from nature by trusting the value of improvisation. Improvisers train for years to be ready for any musical situation, to make something that has never been heard before out of any musical encounter. The nightingale's music is accessible to us humans even though it was not evolved for us.

Even the philosopher Immanuel Kant was on to this. As early as 1790, in the *Critique of Judgment*, he wrote:

> Even bird song, which we cannot bring under any rule of music, seems to contain more freedom and hence to offer more to taste than human song, even when this human song is performed according to all the rules of the art of music, because we tire much sooner of a human song if it is repeated often and for long periods. And yet in this case we probably confuse our participation in the cheerfulness of a favorite little animal with the beauty of its song, for when bird song is imitated very precisely by a human being (as is sometimes done with the nightingale's warble) it strikes our ear as quite tasteless.[2]

So the founder of modern philosophy worried about the art of birdsong that many centuries ago! If we are to value the bird's music, we must get inside his own aesthetic sense. If we are to join in with his music, we cannot copy him, but must learn from him while applying our own humanity to the connection.

Like the nightingale, we are outliers, the extreme results of aesthetic evolution. Who needs our crazy brains, our strange survival strategies, our wholesale transformation of Earth into our weird manufactured habitat, just as the forest does not need the nightingale with its extreme all-night song, and the ocean does not need the humpback whale with its twenty-four-hour song. Listening to nightingales makes me want to repeat myself, since their beautiful utterances are a grand mix of repetition, novelty, change, and silence. A secret code that is not a code, whose meaning is simply the performance itself.

At dawn on one of our first Helsinki days, there are many more birds than nightingales about, especially by the marsh on Tullisaari. Sedge warblers are filling in the mix, and the nightingale strives to cut through. I pull out my iPad tablet, which can produce nearly any sound, and what comes out is pure, resonant, proudly electronic, far from the processes of nature (plate 10). I try to learn from the nightingale, following my own principles as stated above, that manifesto of space, gaps, phrases, rhythms, pauses, contrasts, and clarity. I remember how certain tested bird species such as cardinals preferred synthesized versions of their own sounds to the real utterances of which they themselves are capable. I too yearn for some kind of Platonic musical form out of reality's reach. Maybe that is why we humans also like electronic sounds, or why some of us resolutely dislike them.

I test phrases, bends, and gaps, trying to figure out what to do with this technology. After five minutes I get it: a rhythm is possible, a beat, a steadiness, something the nightingales themselves tend to start and stop—but make it continuous and we've got a groove. The figure has a ground, and we humans feel the bass. At last we can dance. I've brought the biggest Bluetooth speaker I can carry, so these deep frequencies—granted, frequencies the birds can't hear or care much about—are there. But I like to think they can feel them just the way they can feel that an earthquake or seismic shift might be on the horizon.

All right, it's minute seven and the beat is now here. I put down the screen and let it play. I pick up my clarinet: Phrygian mode or whole tones? It's down to me, the bird, and the beat. The corncrakes and crows fall silent, but I feel them begin to dance. There is a clear moment where that nightingale *swings*.

The bottom of the screen in plate 10 shows a regular beat as the iPad wobbles with a software synth called Animoog. Up high are the nightingale's phrases. In the middle the clarinet rises and falls. A soundscape of niches, each one of them filled.

I have had so many students in my college introductory music classes who believe, honestly, that it is only music if it has a beat, if only because so much music makes itself known that way. I get it. You want to feel, you want to move. I wonder how the nightingale feels when his sound is layered above a constant rhythm? Naively, I imagine he must like it, and that is why he and his mate have found such a home among us, in these cities that still have enough greenery to support the wild. I don't even care about arpeggios; the beating modulated tone grabs me. It is enough. I don't like chords, nor do I like harmonic modulation, just frequency modu-

lation to fatten up the tone. To be sure, this machine beat is domi-
nating the wild bird, and I know plenty of people who will hate this
track because it does exactly what I tell musicians not to do: invade
the nightingale's space, overlay his perfection with that incessant
human rhythm. And yet, I cannot deny that to me it sounds cool. I
want to blend with the bird and give him this ground before I go.

Ville Tanttu is filming this for our documentary on music with
nightingales, but he is none too pleased when I pick up the iPad
to join the bird electronically. Clarinet live with nightingale makes
sense. It's pure, visual, and real. But the iPad, this small anony-
mous rectangle, indeed is the very symbol of humanity's tendency
to slot the world into a grid. If you make it a musical instrument,
then it is everything or nothing, a machine of sound with no moving parts that can sample its world and spew back every tone it
grabs, transformed.

Listening back to its results, of course we may find them beautiful. But cinema verité should be made live, and there is just so
much less to see when a musician is out in the field with a crazily
singing bird and he is just moving his fingers on a flat square of
wizardry. To watch it is, quite frankly, boring, like looking at someone playing a video game. It is a dehumanizing, flat-screen version
of the human musical experience. But why is it so much fun to do?
Why does it suck us in?

Obviously, because it gives us such flexibility and power. Anything can be changed into sound; any one sound can be changed
into any other. Since normal human instruments can't capture
the sound world of birds, why not approach these ancient musical
traditions with sounds unexpected and unknown, tones from the
future as much as from the past?

With all this happening I'm forgetting the sun is coming up,
the light is getting better for the camera, and all that background
"birdism" of the dawn chorus is fading out of my consciousness as
I feel the whole piece form. It is the rhythm that guides, the nightingale's start and stop, and the humans' yearning for the beat. We
feel his exuberance all over the pulse.

But what does it *look* like? In a way, there will always be nothing
to see. Electronic sound is an all-knowing domain of limitless pos-

sibility. It is everywhere and nowhere. Global DJ and theorist Jace Clayton, for one, bemoans the fact that everywhere he travels, even remote Berber villages, kids are using the same Fruity Loops and AutoTune techniques. Technology ekes out a sameness that threatens human originality. Or does it? Who knew Animoog was the perfect iPad app to play with cicadas, and that Samplr grabs and shifts nightingale whistles so they can be turned into owl hoots? These tools take music away from what one expects and as such find new life when confronted with the nightingale.

Still, I need to resist the machine and turn to the clarinet to truly feel myself in relation to the bird. The more contemporary the technology, the faster it will become out of date. The more established the musical instrument, the more it can truly extend human initiative to feel like a part of the musician's body and mind. Yet both the new and the old, if we're open to them, quickly enable us to produce things we cannot explain or predict. That's why we make music in these unusual human/animal situations and dare to wrench ourselves out of the familiar.

The whole journey leads to unexpected music, unexpected birds. I cannot leave them out in some singular pursuit of the nightingale's song. I cannot ignore the inscrutable song of the Blyth's reed warbler, an incredible bird with an incongruous name, which makes the song of the nightingale feel rudimentary by comparison.

Every time I come to Helsinki I try to visit the famous music store Digelius, founded by the electronic music pioneer Erkki Kurenniemi and now run by "Emu" Lehtinen, so nicknamed because of his love of birds. And of charging, passionate, unpopular music. Walk in the door and he's never really surprised to see you. Always something interesting is playing. "Isn't it a pleasure," he asks, holding his hands out in the air, "to be able to come to work and listen to Charles Gayle? Or the pied butcherbird!" The great exploratory musicians of Europe and the world might just as easily pass through these doors, and we will run into each other and never really be surprised.

"So you like the songs of birds?" Emu asks me. "Then you really should listen for the Blyth's reed warbler." That was a bird I had never heard of. (see plate 11) I knew of course about all the studies on the sedge warbler, how this is the one bird we're sure of that, when he sings louder, longer, more complicated songs, he does have better luck with the girls and reaches that holy grail of improved mating success. But there is no such luck when it comes to some of these other European warblers with super-complicated songs, such as the marsh warbler, the great reed warbler, and the icterine warbler. They sing, sing, and sing some more, and we do not know why. No correlation with mating of any kind.

But Blyth's? Who was this Blyth? Edward Blyth (1810–1873), curator of the Royal Asiatic Museum of Bengal in Calcutta, got four bird species named after him and the only one without a direct Wikipedia link from his page is the Blyth's reed warbler. "What, Emu, is so special about this bird?"

"Oh, I don't know, David, but I think it's my favorite." Coming from such a master of the world's music, that opinion matters to me. I file the information away. One day I hope to meet a Blyth's reed warbler and hear him sing.

Ville and I drove two hours north from Helsinki wishing to hear some really wild nightingales, birds of the deeper forests, far from the madding crowd of Finns and foreigners and traffic sounds that dominate the nearby suburban island of Lauttasaari. We're up at 3:00 A.M. and the woods are all quiet, not a *satakieli* to be heard. Hardly a sound at all until around 4:00 A.M., when a strange cacophony of experimentation begins to appear, near a little abandoned encampment with rusting barrels and old portable saunas, piles of rusting refuse and low willow thickets. There seem to be three or four birds, ranting and riffing, jumping from one agitated tune to the next. Their music is indescribable, chattering and noisy, but somehow impossible to stop listening to, and, for me, impossible not to join.

It isn't a sedge warbler or a reed warbler. Could it be the elusive marsh warbler, who copies African birdsongs learned on his winter migration? No, they don't make it this far north. Because of the deep music, this has to be a colony of Blyth's reed warblers. A *band* of male Blyth'ses, hardly a female in sight. They don't seem to be competing, but rather engaging in something more like a jam session, singing all together to bring forth the day.

I am reminded of the story of how the dawn chorus sets up a whole environment of marvelous sound, and that birds sing all over the world at dawn, although we still don't know why. This is no ordinary singing bird, but a true composer, an experimenter amid melody and noise, a bridge of sorts between human and avian musical aesthetics.

The iPad agitates him a bit, then spurs him to innovation. He parries with a melody, then leaves space to see what I will do. I am playing the *furulya*, a tiny Bulgarian fipple flute, a bit like a pennywhistle but with two chambers so a drone note can be played, or harmonies in parallel seconds like the Bulgarian State Female Radio and Television Choir perform. The almost-chord is something a bird could do with its syrinx. A few scattered melodic quips and the bird announces a melody that arrests me by sounding immediately human, something rare in the world of bird music, but of especial of interest to those of us looking for a foothold into their world (see plate 12). Dum dum dum, dum dum dum, daaaaah . . . Dum dum dum, dum dum dum, daaaaah . . .

At once, the Blyth's has taught me a melody. With the doubling minor second it is instantly melancholic. It is just a taste, just a glimmer, something hopeful and pure. Perhaps most of you will hear nothing in it. But I love it. I feel vindicated, as if I've suddenly gotten somewhere, entering this musical space between the human and the animal, that ideal place I am always trying to go.

Is this really a Blyth's? How can I be sure? I know nothing about birds, really. But today we have XenoCanto, a crowdsourced website where hundreds of people post their birdsong recordings, and there are many *dumetorum* tracks posted from central Finland. So I'm leaning there. Still, I ask around on Facebook. "Anyone out there know if this is a Blyth's reed warbler?" For a few days, no one has any idea what I'm talking about. But then, corroboration. The great bird recordist of England Geoff Sample chimes in: "Definitely Blyth's. This bird builds sophisticated rhythmic inversions through his development of motifs. These motifs consist of alternating elements: one a percussive salvo of thick, broad-band, and harmonically rich notes, the other a purer-toned, whistled phrase; the combination of the two creates its own internal rhythm.... The Blyth's reed warbler is a master of precise articulation, of balanced phrasing and building thematic variation. There may be a simple formula behind the structure of its song, but if there is, I don't want to know."[3]

I ask him if we couldn't possibly come up with a better name for such an amazingly musical bird. He wonders: "Well, the old Finnish and Estonian birders used to call these little brown critters that just darted around in the thickets 'snake birds'—*madulinnud*—for want of clarification. But we might take a cue from the bird's own sound and call him the *sisichak.*"

*Sisichak, sisichak, sisichak.* I like that. In Finnish the Blyth's is *viitakerttunen*, in Japanese *shiberia yoshikiri*, in Polish *zaroślówka*, and in Faroese he gets his best moniker: *kjarrljómari.* Language is finally beginning to approach music when its diversity is transmuted from babble to repercussion.

Emu, I wish I'd gotten to play this moment for you, since it was you who first told me to look out for this bird. May you rest in peace.


Sound recordists sometimes compress an entire day into one hour, trying to be honest about the progression of different sounds through a twenty-four-hour period even as they speed things up. Messiaen tried to use piano emulations of the species found in a single habitat, to be ecologically honest and respectful to the interrelationships of native species. Some people have slowed down ocelots to turn them into jaguars, or pitched down humpback whales because their actual sound is shriller and higher than we want a whale to be. We trade on impact and stereotypes in field recording, as in the rest of life. They are to be deplored much of the time, but they are also used and exploited.

We are adrift in the park with our machines and our dreams. I can go on and on about what this does to me as a lone human off on my quixotic quest. But I don't want to do that anymore. I am learning to love bringing other people along with me on the journey. It is either an *aha* moment, an *aha* period of life, or simply the sign of growing up. So I decided to write out some suggestions to help others join me on this pursuit:

*Eleven Paths to Animal Music*

1. Forget the name of the bird you hear. It doesn't matter if it's a cardinal (*peo peo peo peo*) or a white-throated sparrow (*old sam peabody peabody peabody*) or a veery (*wheeooo wheeoo wheeoooo*). Just listen to each song as if it's a wonder you have never ever heard before.

2. Leave mostly silence and space. Become just another nameless bird trying to fit your own music into the entire soundscape.

3. If an encounter with the bird music is not changing your own music, then you have not listened long enough. Try to make something new that no one species could make alone.

4. You are not the center of the concert, just one more musician in the mix. Don't feel the need to be in charge. For thousands of years humans have made this mistake. You can change this.

5. Remember, this is some of the oldest music we know. It's millions of years older than our species. There must be something right about it to have lasted this long. Learn from this rightness.

6. Thousands of years of people trying to make sense of what is and isn't musical in birdsong have made it no less elusive. It still lies beyond our comprehension. Work with that ineffability, engage with the other while knowing you will never completely understand him.

7. Scientists may tell us birds care only for the sounds of their own species, ignoring everyone else. That's what they see when looking at what areas light up in their brains as they hear. But as listeners we hear otherwise. These creatures are precisely attuned to sound. All kinds of tunes around them engage their attention.

8. Why do birds sing most at dawn? It happens universally, but even after thousands of years of witnessing this phenomenon we don't know why. We cannot ask other species to explain themselves, since it is not language we share with the birds. It is music. Music does not exist to be decoded. We and the birds exist to make it. Make it together and the whole world feels its power, its joy.

9. Add a groove or a drone with caution. Sure, a repeating, grounding musical force can make any flights of fancy seem logical. But you want to be leery of imposing on a phrase meaning that might not be there.

10. Who can tell what music means anyway, be it human or avian? It's the essence of birds to sing. We are the same. There is so much music in the world, and we cannot escape our yearning for it. We continue to listen and to love.

11. The world needs no more music. It needs no more of us. Still, we keep going on, and the more we listen to everyone else out there, the more we might make music as necessary as what the birds have been singing for millions of years.

# 9

## CELEBRATED BY ALL

For years I made this interspecies music largely on my own, seeing myself as some kind of individual explorer seeking out musical ideas with creatures we can't even talk to. But in recent years I've decided that the point of musical contact with another species is to convince other people to join me. This year when I meet the nightingales of Berlin, I am inviting the best and most adventurous musicians I know to connect in musical collaboration, our species with theirs.

The nightingales are going to help me find the perfect sound. They will guide me there. This time I won't be able to find the path without assembling just the right group of collaborators, kindred spirits, and friends. I and many others dream of a way that humans and nature might live closer together. We all know our species is warming the planet beyond recognition, and that this could mean the end of our reign over this place. Yet there are still these moments during which humans can touch nature through sound happening all around us, as we make music along with the nightingales of Berlin. The paths to animal music sit right before us.

Bernie Krause might call this the "Urban Bird Hypothesis," which would have it that nightingales *like* living among human noises. They haven't only been flung here but have also sought us out. That one bird still perches atop the streetlight at the loudest corner in Treptow, where Eisenstraße meets Puschkinallee, near the S-Bahn station for the park where most of his compatriots are. He is committed to this territory, and either there are some urbanized female birds nearby as well or perhaps he is destined to fail again and again. Maybe that makes his song better and better.

I agree with Krause when he stresses that humans are destroying the vast richness of natural soundscapes, but at the same time we create new, interspecies soundscapes. I would never say these are better than what the wild has offered us, but they do seem inevitable. The wild may be perfect, eternal, endless, deep, and serious. But we are just little flighty, edgy, nervous, flitting human birds who tinker and transform, making desperate forays into contact with the eternal wild world, and in the city this can happen in a newly energized nature. The nightingale sings "Ich bin ein Berliner," and we can do more than smile. We must join in.

I love to listen to different musicians respond to the song of the nightingale for the very first time. I have played with these birds for several years now, and sometimes I wonder why I keep trying to make music with musicians with whom I cannot speak, who as birds live such different lives from people who could join the band. Some critics think it's all delusion, that I intrude upon the birds' ancient world of perfect sound and struggle; but whenever I bring a new musician along to play with the nightingales in Berlin, I realize why I began this process in the first place. We all feel such joy and hope when music can carry meaning across species lines.

Not every human musician fits into such a scene, but I have found special people who do. First up is Korhan Erel, who is never afraid of projects that are difficult to explain. He understood right away that playing music with nightingales presents quite a challenge, because their music follows rules and criteria we can hardly know.

Erel calls himself not a nightingale but a "fightingale." I roped him into making music with nightingales when I heard his remarkable electronic improvising abilities at the tiny Neukölln club Sowieso. He has moved over the last few years in Berlin from Istanbul, a journey that is not always easy. Germany may be famously open to refugees but is still suspicious of unclassifiable musicians wanting to settle permanently in Berlin.

Erel has a sense of balance and organization different from that of many of the city's experimental musicians, quite possibly be-

FIG. 13. Korhan Erel

cause he spent a few years working in the corporate world and knows he does not want to go back. Yet that experience gives him a certain practicality that can follow you everywhere. He organizes concert tours, festivals, and events, knows how to stay committed and put them together. His musical reputation spreads and expands. He is always willing to take on new and unprecedented projects, including a trio called the Liz, every member of which is called "Liz," two women and one man (Korhan) who dress up as ancient Egyptian goddesses and perform a vast neo-Pharaonic ritual called the Book of Birds. Here's why Korhan is so interested in the nightingales' responses:

> I find some of the nightingales' songs inspiring: their use of rhythm, the sudden changes, the jump from one thing to something totally different, their use of tempo and speed. I think they're extremely musical, not what we're used to hearing. Their songs are not melodic, but might even be algorithmic music in a way, like generative music produced by algorithms, because the changes are sometimes so abrupt. Most of the music I play is perceived as chaotic by other people, so I guess that makes me a nightingale as well. People don't always get it. More often they think my music is random.

Erel has spent years perfecting a way to play computerized sounds using multi-touch-sensitive controllers developed by Keith McMillen called QuNeo, an iPad-sized grid of sixteen gel pads with five or six dimensions of touch sensitivity to each. One can program each to be an extremely sensitive and nuanced controller of electronic sounds. When Erel plays this controller connected to his computer, loaded with synths and samples and effects, he calls the whole integrated device the Omnibus. It is a mysterious and personal technological situation that only he understands. It evolves from year to year, and by now he's probably using a completely different setup. He composes and improvises along with other people, as any instrument can be added to an ensemble. At times he samples the sounds other musicians are playing, at others brings only his own sounds. In a way, playing with him is like playing with a nightingale, since one hears the order in the sounds without knowing exactly what is happening.

The album of duets we made, released in 2015 as *Berlin Bülbül*, is something of a prequel to the Nightingales in Berlin project, which is now being filmed. *Bülbül* means "nightingale" in many Middle Eastern languages, and this duo project showed the possibility that interesting music could be made in tandem with nightingales or in emulation of the way they work together among themselves. We took this project to the Borusan Culture and Arts Center in Istanbul, where our audience applauded the confluence of cultures and migration of sound.

When Erel is out playing live with the birds he brings no pre-recorded sounds. The whole performance is developed out of direct sampling of live birds, transformed into new cut-and-pasted music as he plays along. He is the instigator of this approach, which I have learned from him and sometimes done on my own, along with my clarinet, as in the Helsinki recordings. I feel more comfortable having him do it, because he knows that language and is closer to inhabiting the machine as an instrument, a clear extension of his own musical ideas that try to evade the expected and the familiar. He explains why this swirling outreach sometimes works well with my clarinet:

We don't play very densely. I don't use really avant-garde sounds, and our duo does not sound very abstract. Which I like. I think your clarinet breaks the abstraction, it's a good mix of the weird and the not so weird. It keeps me in line. I could play really crazily, but I don't. Of course, people who are expecting dance music when they hear I will be electronic might be disappointed, because they can't really deal with it. But our music strikes a good balance.

The Omnibus works in the studio, and when playing clarinet with it I think of emulating the way nightingales play with each other. Our studio pieces have no real bird or bird samples in the mix, but they are clearly influenced by the way we play in the field with nightingales. Their musicality has changed us, and we won't change back. In the field Erel plays with an iPad empty of sound using an intuitive app called Samplr with no words, no visible instructions, shaping sound in ways you learn only by practice and routine, like any profound musical instrument. He one-ups the bird by bouncing the bird's own ideas back to him, twisted and transformed.

The first document of our connection appears on *Berlin Bül-bül* as "Dark with Birds and Frogs," a recording from our first live concert with a bird in May of 2014, the anniversary of the night the war ended, that night in chapter 1. This is us along with the very same bird, one year after we found ourselves chastised by the scientists. One special bird, who every year comes back to the same tree. This bird appears on Silke Kipper's map of nightingale territories, where she names him "number 7," and it is him with whom I also performed with Lucie Vítková in 2015. Now I still worry about upsetting Kipper's research and compromising this bird, but this is Berlin—he has already heard so much. Back then his tree was in front of an old rock-rimmed frog pond, since demolished in the "renovation" of Treptower Park. Not sure exactly what tree our boy will come back to this year, but I don't think a little construction project will be enough to deter him.

On this 2014 recording, done in front of a live audience at midnight, the nightingale dances musically in and around samples and transpositions of himself, and just after he releases the infamous buzz, the frogs are enlivened into action. They, too, are sampled by Erel, and the whole environment of sound becomes raw material for a music that, in retrospect, we can't exactly parse. I like that. But far be it from me to know if this music we make with birds is any good.

"So, Korhan, tell me about that one special sound we now hear the bird making."

"That was the *boori* sound. This is when the magic happens. This makes the girls crazy. If I play this sound to my wife she'll say, 'Are you kidding me?'"

There is that familiar rasp from Kipper's rigorous nightingale buzz song paper. The *boori* sound is right there.

"Is this an ugly sound?"

"Pretty, ugly, these are all human inventions. Until we can find a way of speaking to the nightingales, everything we conclude about them is bullshit! Sorry, scientists. I don't think we need to find conclusions; we should just enjoy their songs. We're getting more material from the bird than we are giving to him.

"He doesn't need us, since his kind have been around for millions of years, doing just fine."

"Berliners might like this music, but they tend to frequent clubs, not parks, in the middle of the night. Although one is rarely alone in Treptower or Hasenheide ..."

Korhan Erel speaks with certainty, freedom, and a dash of petulance. I ask him where he wants to be in five years, and he says, "By then I hope to have my own festival." Then, half-facetiously: "I'm still not famous enough." This fightingale fights back. About his music, about his politics, about his right to wear makeup and dress as Liz Erel if he wants to. Berlin is his city, exactly where he wants to be. There is no better musical partner to have beside me when performing with a nightingale, because he plays an instrument impossible to define or explain. He makes machines come alive under his fingertips, unafraid of sounds that he has never heard before. He's willing to learn from a bird who has been twisting sound around for more years than our kind has been in existence.

One needs awe and respect to do this, but also courage enough to sound one's ground.

And he doesn't always need machines. He's the kind of guy who can whistle a silent nightingale out of the bush into song. When it seemed as though we heard nothing one cold April night, a hundred people already disappointed and waiting, he got our bird back out on a limb. The concert was able to happen. Sometimes imitation is the highest form of avian flattery.

I must apologize for quoting so much European poetry thus far in this book, when it is in Persia that the word "nightingale" has always been an honorific to name the great musicians. People live through poetry even in today's Iran, though the government tries so hard to keep them down. Our bird remains the symbol of the depth of love that pervades so many of that nation's verses:

> Nightingale!
> Enjoy the rosiness of the Beloved's cheeks,
> Your song is celebrated by all.
>
> The Beloved's tender lips cure all ills,
> The Beloved's treasure-house is filled with rubies.
> My body may not always have you near,
> But my heart remains the gateway to your resting place.
>
> My heart would never permit a cheaper love to enter it,
> The treasure chest of my heart is sealed with your magic name.
>
> Since the world loves the Beloved, I must Love the world.

Hafez, remarkable how you have always been the most popular poet in the world, for hundreds of years, even in English. Your culture embraces the sheer power of poetry. You are not scared to let loose, to set off that sensual power like the song of the nightingale, so foreign yet so close to us. He is eternal, his energy limitless. The rose of his obsession will wilt and die, and can prick us at any time. A fickle choice for love, an irresistible pull. The nightingale is an essential image in Persian poetry, for its relentless love, intractable passion, and a commitment we humans can only dream of touching.

Yet Hafez also praises drunkenness, debauchery, pleasure itself, something we are too often told to be embarrassed by. The nightingales in his poems know how to party:

> The nightingales are drunk,
> wine-red roses appear,
> Sufis, all around us, happiness is here;
> How firmly, like a rock, repentance stood.
> Look how a wine glass taps it—
> and it lies in pieces now.[1]

Cymin Samawatie tells me there was a time in Berlin when those approaching death would ask to be carried out into the streets at night to hear a nightingale sing for them one last time. The power of this bird's hundred songs cuts through anything we humans may try to do with the world, gets louder the more we fill its environment with our own noise. It is the living symbol of a nature tough enough to stand up to our human wiles, our idiocy, for we can long to know it but will never completely get there.

Samawatie is one of the most remarkable musicians in Berlin. A singer well versed in both Persian odes and jazz standards, she grew up in Germany with parents born in Iran who made sure she knew the best of both cultures. Her lyrics may come out in Farsi, Arabic, English, German, or Hebrew, depending on the time and place. She is committed to doing it all her own way. If she ever returns to her home country, she will be praised as a nightingale herself.

She is one of few artists who have walked right into the headquarters of ECM records to announce to its legendary leader, Manfred Eicher: "People say I should record for you." Since then she has put out three albums on the famous label with her band, Cyminology.

I've heard her sing in different contexts, different stages, different parks, always beautiful, always firm. Treptower. Viktoriapark. Gleisdreieck. She is not even allowed to sing in Iran, her native

FIG. 14. Cymin Samawatie

land, where women are not allowed to sing for men, or anywhere in public. People there listen to her in secret, though. They know that intercultural beauty is possible. Nations want to decide what is allowed and forbidden, but they cannot stop the nightingale from flying easily across borders. They cannot stop humans who want to join with nightingales. Hafez, drunk on nightingales and wine, continues:

> If there's no sorrow
> there can be no happiness,
> When the world was made
> men knew this and said—Yes.

> Rejoice, don't fret at
> Being and Non-Being,
> Say that all perfection
> will be nothingness some day.

I have helped scientists try to calculate the meaning of the nightingale's song, all the while knowing such a task to be folly. Still, if science doesn't try to figure it out, what kind of power does science really have? *Why* is this most ultimate of bird musics sung most surely at night? *How* will we know what the best song of this creature will be? Each human tongue's different way of trying to represent this bird shows a distinct onomatopoetic way of picking a glimmer of its order and delight.

I won't be afraid to enumerate and analyze his phrases, and I will sample him on my little screen and remix my slices of his power into simplified beats that humans can hope to understand. Still, I'd rather hear Samawatie sing the words of Saadi extracting honor through the leaves:

> I saw *bülbüls* commencing to lament on the trees,
> the frogs in the water and the beasts in the desert
> so I bethought myself that it would not be becoming
> for me to sleep in carelessness
> while they all were praising God.

> Yesterday at dawn a bird lamented,
> Depriving me of sense, patience, strength and consciousness.
> One of my intimate friends who
> Had perhaps heard my distressed voice
> Said: "I could not believe that thou
> Wouldst be so dazed by a bird's cry."
> I replied: "It is not becoming to humanity
> That I should be silent when birds chant praises."

That's the nightingale resounding over the centuries, an invocation and a puzzle, the very root of music and of life itself. We are not meant to remain mute in the face of this most beautiful music of the birds. But how exactly not to be silent when the *bülbüls* chant praises to the world? The rest of them resound together at dawn, after the nightingale has been priming us in his loneliness all night.

Out in Gleisdreieck Park at midnight, Samawatie softly sings a few more of Saadi's ancient words:

> O nightingale
> you are in love with a flower—
> I am in love with a man.

I pick up my clarinet, trying once again to become a bird. It might actually be easier than playing human music, since these songs of evolution have coursed through our DNA for millions of years before our species ever defined itself on this planet. These futuristic electronic birdsongs come deep out of our evolved past. Samawatie sings on:

> When did kindness end? What brought
> The sweetness of our town to naught?

A soft ensemble of humans tries to cross the species line with reverence yet with growing confidence. Berlin has room for all of this, at any time of day or night, in the spring, when the forests are alive with song.

> The ball of generosity
> lies on the field for all to see—
> No rider comes to strike it.

More of us should go there, beyond the myth of nightingale killing himself through love with the thorn of a rose. It's not enough to touch the song of the nightingale only when it's time to go. The extremes of the moment and the sound are able to make us more alive. That's what Hafez is getting at when he thinks the nightingales are drunk, what Saadi knows when he cannot stop himself from joining in. I wonder what music he sang to them?

I once heard a song in a downtown New York concert, a beautiful piece with lots of silence. The Estonian-born singer Lembe

Lokk lives outside Paris in a town by the River Seine. She has published poems in French, English, and Estonian and has performed throughout her life in different personae: Madame Rouge, Lemmbe, Gjaàr. She always looks between identities, cultures, countries, and kinds of music. Something Nordic but not really. The moment I heard her sing this song, I knew I wanted it sung in Berlin at night, along with nightingales:

Dreaming slow
Clear voices I can hear
Through the hasty words of
farewell

My heart
is still

Dark storm is turning white
The fierce renewing wind
will blow

Dreaming slow
The winter needs me now
The nights are long, the healing
so rare

I'm running through the air
The morning will need me
to go

Dreaming slow
My mind is sitting there
My icy breathing traces
on the snow
a flame

The fireworks of spring
The mirror of words
to be shown

FIG. 15. Lembe Lokk

I asked Lembe what this song is all about.

"I wrote it in Paris, ten years ago, for a session in Berlin. I was thinking about all this climate change, the wind, the temperature, the changing Earth. Every time I sing it, it takes me back home to Estonia, the cold frosty spaces. I see all these Estonian winter landscapes with a lot of ice and snow. Once I was touring in Estonia and we crossed the sea the very last moment they had the highway open across the ice. In Estonia in winter we drive across the frozen sea. Here you might not believe it but we do, and the day we crossed the sea was the moment they closed the ice road. I saw huge blocks of ice. The road was about to crack, but we made it."

That started to make a sense, I nodded. "This song has an alluring space between each phrase and the next. What was it like to fill that space with the songs of nightingales?"

"It made me fly, in a way. It's the first time I sang with birds in general, with nightingales in particular. In the end you really feel like you are flying. It's so strange to try listening to them, up there, so you tend to this song, and at the end, when you get really into it, you feel like flying, actually."

"What does it mean to 'dream slow'?"

"Can you dream fast?"

Do the nightingales care? Do they listen? Do they stay themselves, or change as humans change to dare making music with them? Saadi implores us not to lie mute in the face of their beautiful midnight songs: *How can we be silent when birds chant praises?*

Saadi is saying he must be a poet in the face of such birdly devotion. But we, all of us nightingale musicians, take the bird as our muse and our partner, and we strive to find a way to join in with him. I have written about this regarding different creatures—birds, whales, bugs—for years, and sometimes it seems as if I envision myself alone with them, but it is never so. Neither should it be so. This music with nature becomes powerful when we bring others along with us, when the music rouses an audience, shows them a beautiful possibility in the here and now, the hear and now, listenable all around us.

Viktoriapark is one of Berlin's most exotic parks, boasting the biggest artificial waterfall in the city, cascading down wildly from the slopes of the Kreuzberg. In its tree-filled upper slopes live some of the most virtuosic nightingales in the city, and we were lucky to find one on a surprisingly cold early May night. The Finnish violinist Sanna Salmenkallio listened carefully to his whistles and bowed smoothly along as he surprisingly changed his pitch along with the continuous tones. The violin, when played so expertly, is probably a far better instrument than the clarinet to swing along with these birds.

After an initial round of phrase, Lembe joins in with violinlike gasps of wordless voice. Our *Nachtigall* is fervent and pure, relentless and firm. I am the third to join in, flummoxed by all this sonic possibility, pushing my own shrieks to their animal limit. Once again I wish and want to be a bird. After all that release I settle in to a low minor arpeggiated search. For a melody, for a home.

At first I doubt that all three of us can join with him at once. When we finally agree to this, we are all extended beyond the sweetness into gasps and howls, ultimately racing for human-bird release. With this music, we all have the right to be extreme. He screeches with a *boori* to set things right. Who needs anyone to win here? Our bird has picked a perch atop the forest on the hill looking down over the whole city of Berlin, dreaming slowly on of this inscrutable song.

Is our human sense of musicality enhanced or insulted when we try to join in with these noble birds? The best cities do contain acres of greenery and are full of nightingales, seemingly more of them than in the pristine wilderness. These may be birds who like liminal habitats, borders between forest and field, a kind of landscape much more common when made by humans than found in the wild. It may be that, inadvertently, human cultivation of the land has created ideal habitats for this bird.

This idea gives me hope. In some places we may have inadvertently improved the world just a little bit for nightingales, and if we now remember to listen we might value this accidental landscaping we have done. If humans can live on this planet in a way that keeps the nightingale singing and happy, we have tended our garden well.

Nightingales are far better known as symbol and idea than through their songs, which sound so real. And what better way to pay tribute to this song—so often celebrated, but too often cerebrally—than to find a way to join in? I know what qualities my music has developed in order to make it possible to fit into birdsong, whale song, and bug song, but now it is through seeking out others' voices that the whole project heads for greater depth.

I started writing this book with a mind to expand from interaction with nightingales to the whole problem of how human music fits into the music of nature. I wonder if I have gotten it right, this contrast between the first encounter of a new musician with this ancient birdsong and my own excitement at learning so much from so many new people. So many hours of new, unplaceable music, blending one species and another. I imagine handing it to Tina Roeske to feed laboriously through her algorithms in order to come up with statistical visualizations explaining what it all means. At least one scientist dares speak of beauty and is wise enough to admit she can't explain it when she says, "I'm worried I won't be able to get any funding to continue such uncertainty." The lives of artists and scientists might not be so different after all.

Repetition, repeat on repeat, over and over again, over and out. The nightingale's song is ever new and always the same. I'll always know it's a nightingale, but is it a thrush nightingale or a common nightingale, an *urretxindor* or a *fülemüle*? Choose your own language to find the best word.

At least I know, just as when a few of us human musicians are out there parlaying with these ebullient birds, that the nightingale and his song will outlast us, simmering on all through the night as he has for millions of years, whether or not there our human cities were there to welcome him. One, two, three hours at most is what we humans can take, playing out in the dark with his whistles, clicks, and *boori*s. When our hands get too cold and we can no longer see the melodies for the trees, we all will pack up and go as the *bülbül* sings his victory. If it's a battle, then the bird will always win. If it's a debate, then he'll always have the last word, soloing on until the frenzied dawn brings his cadenza to a close with that resounding symphonic mystery, the daily conference of the birds.

In Peter Handke's novel *The Moravian Night*, the protagonist, a seasoned "former" European writer, attends a conference on the omnipresence of noise in modern life. This gathering is full of people for whom noise has made life completely miserable. As he often does, Handke turns contemporary despair, about one thing or another, into something poignant and beautiful:

I've dreamed about Original Sounds, plural, one and then another. I not only dreamed about them but heard them in broad daylight, and I was perhaps never more awake. And I have heard the Original Sounds as a voice. Yes, the flapping of a butterfly's wing, even if by now it's nothing more than that: at one time it tuned me in, once upon a time at least. Or the splash of that one dewdrop long ago, a splash on the margin of audibility, and then, at an interval, once I was tuned into it, another drop at the same great interval, landing let's say on a piece of firewood, on pebbles in the gutter, on the sidewalk, always in the same spot, until I heard the dewdrops as the regular tick-tock of an unheard-of clock, which at the same time became audible in the inner ear. . . .

Original Sounds used to mean impressions that would remain in one's ear forever—or so they promised. In one's ear? No, in one's heart, where they had first echoed, corresponding to the Original Sound before all Original Sounds—so similar after all to that voice in a dream that once woke me as I have not been wakened before or since. . . . At times I even heard a murmuring, and similarly a certain whispering. And a certain hammering? Yes, that too. And one or another booming, roaring, rushing, shrilling, drumming? Well, why not. . . .

In the current noise I have come to close to losing my soul. The most destructive thing about this noise is that against my better instincts I am forced to identify the noise-generators with their noise . . . A single lovable sound, and my soul will be healed. Secrecy—show me the place where you are hidden.[2]

The original sound, the perfect sound, the nightingale's peak, the sharawaji place—it's here, it's there, wherever you dare to listen. I wish I could tell you it is right around you wherever you are. I *can* tell you that, but I know that may not be enough. Keep attending, keep on moving. You will get there one day.

In 2016 the world lost an animal sound forever when its last Rabbs' tree frog died in the Atlanta Zoo. No more remain in the wild. The call is short, raspy, matter-of-fact: like any frog, but then again not. It's a sound we shall never hear again. It repeats to form a beat, the most basic kind of music in the world. I duplicate it, mirror it, echo it, surround it and swirl it with a sonic image of an imaginary chorus of hundreds of these frogs, singing around us and so happily. It's a pure fantasy, as there are no more Rabbs' tree frogs, and we continue to lose two hundred species every day. (Take heart, however, as we still discover about forty-three species each day.) Perhaps I am foolish to stay optimistic, but how else to live in the throes of never-ending bad news? More and more losses like this will occur as the century goes on.

I mixed the Rabbs' tree frog into a new piece. The group sound dissolves again to the lonely one, and this morphs into a single snowy tree cricket, a hard-to-see but not altogether rare insect whose song is familiar, rhythmic, and utterly inspiring. The music goes on, a bed of happiness. As it fades into the background, the tremendous bugling of an elk joins in, as seen in the sonogram in figure 8, a recording of a captive herd in Pennsylvania by the great Lang Elliott. The confident bellowing then segues to the majestic song of the humpback whale, that greatest and most enigmatic of animal singers, sometimes going on for twenty-four hours straight, and at the end this is whipped up into the song of the thrush nightingale, so very like a whale but slipping by ever so fast. What holds the whale and nightingale song together? Show me the place where you are hidden; something about the music of each of these real outliers of evolution connects, and we do not know why. Perhaps we should never know why.

As the summer draws to a close, the nightingales of Berlin decide it is now time to go. One lone nightingale is still chanting late in summer on a chilly day as the leaves start to fall around the grand Russian war monument in Treptower Park. It's too late in the season for him to be singing, but sing he does, oblivious to what he is supposed to be doing or what the rules of nature have planned for him. A wind rises from the north pushing south, and he knows it's time to go. He alights and begins the journey. We see him traversing the air.

He crosses toward Prague, on to the Middle East. He crosses the Arab lands whose poets have for centuries praised him. Our bird keeps going, flits and swoops all the way down to Ethiopia—a high, dry forest—where the same musicians we saw at the beginning are huddled over their covered instruments, braving a sandstorm with shawls over their heads. The storm dies down, the desert minstrels uncover their instruments. They are not rebabs or sintirs, but DJ controllers and 16-pad grid samplers, the kind any electronic musician might use today anywhere in the world.

The others uncover their heads. They are our familiar musical travelers: Cymin, Korhan, Lembe, David. They are ready to begin. The bird starts to sing, and we cannot help but join in. We too want to be the opposite of time as the whole story starts all over again.

I look for words absent from the whole history of nightingale poems, words that no one else has said. If I find those words I will finally know what it is I need to say.

Nightingale—
Your song outlasts us all.
You've sung it long before we ever counted time
or tried to crack your midnight codes.

Always the same or never the same—
Who cares! Cioran said humans are the only apes who
worry we might get bored;
how happy that gorilla is staring into space day after day,
Are we really better off than he, with our destruction and our
      fears?

Nightingale—
Your song has been here long before we first arrived.
Must we ask forgiveness before we join in?
We've misread you for years ... hearing madness, unfulfillable
      love,
Your hymn is more sincere than that.
Compelled to sing, you
always know just what to do.

When are humans ever so sure—of anything?

As long as we don't wipe out your world,
and don't completely figure it out,
you're never going to stop.

That's right my *bülbül* friend, *please* don't stop.

Never end.

# ACKNOWLEDGMENTS

It has been five years since I first went to Berlin to make music together with nightingales. Thanks to an invitation from Reinhard Schäfertons, I was able to be a guest professor at the University of the Arts. Thanks to all my students there, and to my friends and collaborators in Berlin: Bernhard Wöstheinrich and Christine Kriegerowski, Markus Reuter, Tobias Fischer, Korhan and Tuçe Erel, Cymin Samawatie, Ralf Schwarz, Ari Benjamin Myers, Andrea Parkins, Justin Lépany, Frédéric L'Epée, Lucie Vítková, Lima Vafadar, Reelika Ramot, Jessica and Martin Ullrich, Silke Kipper, Sarah Kiefer, Constance Scharff, Michael Obert, Alex Tondowski, Chen Yang, Robert Henke, Maria Magdalena Wiesmaier, Shin-Hyang Yun, Tomas Saraceno, Bernd Brunner, Lars Schmidt, Zabriskie Books, Jensus, Christina Wheeler, David Abravanel, Brigid Gilbert, Todd Burns, Jeffrey Goldberg, Peter Cusack, Gaëlle Kreens, Erika Hoffmann, and Hans Peter Kuhn.

Over the years many others have encouraged my bird-musicking obsessions or helped this project along in other ways: Michael Pestel, Laurie Anderson, Marilyn Crispell, John Wieczorek, Robert Jürjendal, Jaron Lanier, Matthew Aidekman, Gunhild Seim, Benedicte Maurseth, Nicola Hein, Hans Tammen, Max Eastley, Richard Prum, Alvin Curran, Olga Mink, Helen Arusoo, Alexandra Duvekot, Sam Auinger, Lasse-Marc Riek, Martin Pedanik, Daniel Ladinsky, Gilles Alvarez, Hollis Taylor, Ofer Tchernichovski, Pauline Oliveros, Timothy Hill, Iva Bittová, Tina Roeske, Kate Rigby, Sam Lee, Kerry Andrew, Marcus Coates, Mark Pilkington, Bernd Herzegonrath, Alexander Pschera, Anna Roberts-Gewalt, Tim Dee, Francesca Mackenney, Ville Tanttu, Mete Sasioglu, Petri

Kuljuntausta, Dario Martinelli, Katja Hägelstam, Rauno and Outi Lauhakangas, Rachel Mundy, Edward Hirsch, Edie Meidav, Joan Maloof, Richard Powers, Lawrence Weschler, and Brian Dolphin, to name a few.

Thanks to my agent Markus Hoffmann, editors Christie Henry, Marta Tonegutti, and Alan Thomas, and editorial associate Susannah Engstrom at the University of Chicago Press for shepherding this project to print. Thanks to the New Jersey Institute of Technology, Provost Fadi Deek, Dean Kevin Belfield, and chair of the Department of Humanities Eric Katz for turning me loose on a sabbatical year to Berlin. I hope they are pleased with the results.

To my closest advance readers, John P. O'Grady and Evan Eisenberg, whom I first trusted to assess my words: glad that you both live in the mountains close by. Thanks for expert editing by Tyran Grillo, a keen writer on musical matters as well. And thanks to my wife, Jaanika Peerna, a wonderful artist and understanding partner, and to our son, Umru, a sonic explorer in his own right who will be remixing nightingale sounds soon enough.

And to the birds, who will outsing us all.

# FIGURES

FIGURE 1. David Rothenberg playing to imaginary whales in Svalbard. Photograph by Andrea Galvani, courtesy of David Rothenberg.

FIGURE 2. The sexiest of all nightingale tones, the *boori* sound

FIGURE 3. Olavi Sotavalta's structural analysis of the song of the thrush nightingale

FIGURE 4. Stereotyped versus variable performances by four different nightingales

FIGURE 5. The spaces between each nightingale syllable analyzed in a single picture. Diagram from David Rothenberg, Tina Roeske, Henning Voss, Mark Naguib, and Ofer Tchernichovski, "Investigation of Musicality in Birdsong," *Hearing Research* 308 (2014): 71–83.

FIGURE 6. A birdsong chorus above the thrum of a passing plane. Diagram by Almo Farina from N. Pieretti, A. Farina, and D. Morri, "A New Methodology to Infer the Singing Activity of an Avian Community: The Acoustic Complexity Index (ACI)," *Ecological Indicators* 11 (2011): 868–73.

FIGURE 7. The acoustic complexity index in pristine versus noisy soundscapes. Diagram by Almo Farina from Almo Farina, Nadia Pieretti and Rachele Malavasi, "Patterns and Dynamics of (Bird) Soundscapes: A Biosemiotic Interpretation," *Semiotica* 198 (2014): 241–55.

FIGURE 8. Bugling elk in Pennsylvania, recorded by Lang Elliott

FIGURE 9. A mockingbird duets with a police siren

## *Color Plates*

ization of Long-Duration Acoustic Recordings of the Environment," *Procedia Computer Science* 29 (2014): 703–12.

PLATE 7. Bernie Krause's "Great Animal Orchestra" installation in Paris

PLATE 8. The most beautiful sound in the world? A Borneo soundscape

PLATE 9. "Sharawaji Blues," early morning in Helsinki, clarinet and nightingale

PLATE 10. "Alien Beauty," four minutes in, iPad and nightingale

PLATE 11. Blyth's reed warbler. Film still by Ville Tanttu, from the documentary *Nightingales in Berlin*.

PLATE 12. The Blyth's reed warbler teaches me a tune

PLATE 13. Nightingale singing in Volkspark Hasenheide. Film still by Ville Tanttu, from the documentary *Nightingales in Berlin*.

PLATE 14. Where we found the birds

# FOR FURTHER READING

There are only a handful of books written entirely on nightingales. The earliest I found is Oliver Pike, *The Nightingale: Its Story and Song* (London: Arrowsmith, 1932). Richard Mabey wrote two excellent ones in recent years, *The Book of Nightingales* (London: Sinclair-Stevenson, 1997), and *The Barley Bird: Notes on the Suffolk Nightingale* (Framingham: Full Circle Editions, 2010). Edward Hirsch assembled a masterful collection of all the poems about nightingales he could find, *To a Nightingale: Sonnets and Poems from Sappho to Borges* (New York: George Braziller, 2007). One of the best long chapters on nightingales is in Ton Lemaire, *Op vleugels van de ziel* (Amsterdam: Ambo, 2007). Yes, it's in Dutch, but you'll figure it out.

On birds and music in general, an extraordinary recent book is Hollis Taylor, *Is Birdsong Music?* (Bloomington: University of Indiana Press, 2017), by far the most detailed work on this subject so close to my heart. My own book *Why Birds Sing* (New York: Basic Books, 2005) discusses the music, poetry, and science of what humans know about avian aesthetics, with reference to many more classic texts than this present volume. Since then a few new relevant gems have come out, such as Donald Kroodsma, *The Singing Life of Birds* (Boston: Houghton Mifflin, 2005), and Johan Bolhuis and Martin Everaert, eds., *Birdsong, Speech, and Language* (Cambridge, MA: MIT Press, 2013).

Everyone has their favorite books on Berlin, one of the most written-about cities in the world. I like Joseph Roth, *What I Saw: Reports from Berlin, 1920–1933* (New York: Norton, 2002); Chloe Aridjis, *Book of Clouds* (New York: Grove Press, 2009); Anna

Funder, *Stasiland* (New York: Harper, 2011); and Paul Beatty's hilarious *Slumberland* (New York: Bloomsbury, 2009).

On the study of ecoacoustics there are few detailed books; probably the most comprehensive is Almo Farina, *Soundscape Ecology* (New York: Springer, 2014). The only book that mentions the sharawaji effect is Jean-François Augoyard and Henri Torgue, *Sonic Experience: A Guide to Everyday Sounds* (Kingston, Ont.: McGill-Queens University Press, 2006). Brian Kane's *Sound Unseen* (New York: Oxford University Press, 2014) delves into the value of strange sounds that we cannot name. Bernie Krause has written the excellent *Great Animal Orchestra* (New York: Little, Brown, 2012) and the more recent *Voices from the Wild: Animal Songs, Human Din, and the Call to Save Natural Soundscapes* (New Haven, CT: Yale University Press, 2015). The beautiful bilingual catalog of his exhibition at the Fondation Cartier, *Le grand orchestre des animaux* (London: Thames & Hudson, 2017) is spectacular. Gordon Hempton searched for *One Square Inch of Silence* (New York: Free Press, 2009), and his newer and instructive *Earth Is a Solar Powered Jukebox* (Port Townsend, WA: Quiet Planet, 2016) is available only from his website, quietplanet.com. Lang Elliott has authored many excellent books, some with his pristine recordings included, such as *Music of the Birds* (Boston: Houghton Mifflin, 1999) and *Guide to Wildlife Sounds* (Lanham, MD: Stackpole Books, 2005).

*Nightingales in Berlin* has learned much from the world's poetry that celebrates the nightingale. I have been inspired not only by Ed Hirsch's fine anthology, mentioned above, but also by Simon Armitage and Tim Dee, *The Poetry of Birds* (London: Penguin, 2011); Billy Collins, ed., *Bright Wings: An Illustrated Anthology of Poems about Birds* (New York: Columbia University Press, 2009); Attar, *The Conference of the Birds*, translated by Sholeh Wolpe (New York: Norton, 2017); Graeme Gibson, *A Bedside Book of Birds* (New York: Talese, 2005); and of course all the fabulous poetry by Hafez, found in so many fine translations, including the creative ones by Daniel Ladinsky such as *The Gift* (New York: Penguin, 1999) and *I Heard God Laughing* (New York: Penguin, 2006), and the rhapsodic versions of Dick Davis, *Faces of Love:*

*Hafez and the Poets of Shiraz* (New York: Penguin, 2013), and his more recent *The Nightingales Are Drunk* (London: Penguin Classics, 2015). The words of Saadi are harder to find in English, but Wheeler Thackston translated the *Gulistan* (Bethesda, MD: Ibex, 2007), and other versions can be easily found online.

And why are nightingale songs so beautiful? Sexual selection has something to do with it, a theory of Darwin's given short shrift for a century until these three recent books tried to tackle it: my own *Survival of the Beautiful* (New York: Bloomsbury, 2011); Richard Prum, *The Evolution of Beauty* (New York: Dutton, 2017); and Michael Ryan, *A Taste for the Beautiful* (Princeton, NJ: Princeton University Press, 2018).

# *NIGHTINGALES IN BERLIN*: THE MUSIC

This book has a dedicated website, www.nightingalesinberlin.com, which includes links to audio examples for all the sonograms that appear in the text, photographs of many of the musicians who play along with me and the birds, and various videos and more sounds that augment the story told here. Links to all the sonogram examples that appear in this book also appear on Soundcloud.

I have released two previous albums of live music with nightingales, a single trio performance with Lucie Vítková on voice, myself on clarinet, and one excellent bird called *And Vex the Nightingale* in 2015, and *Berlin Bülbül* with Korhan Erel on electronics, containing some live performances from 2014 in the parks of Berlin.

The new album *Nightingales in Berlin* released along with this book is composed of live recordings made in 2016, 2017, and 2018 in Berlin and Finland. These tracks listed below are can be found in download and streaming form at the usual online sources. A double CD entitled *Nightingale Cities* is also available, including tracks from both Berlin and Helsinki that are not to be found online.

1. The *Boori* Sound (3:35)
   Viktoriapark, Kreuzberg, Berlin, May 5, 2017
   Lembe Lokk, voice
   Sanna Salmenkallio, violin
   David Rothenberg, clarinet
   Nightingale

There it is, that sexiest sound, barely there, at 0:33 and 1:52, heard over the recurring minimalist cycle of human voices and instruments grounding the song of the nightingale, which precedes all human music. This piece epitomizes what is special about this edition of interspecies music making—it is a group process, not just me alone with the birds, but more of us. The ensemble grows.

2. Dreaming Slow (7:32)
   Volkspark Hasenheide, April 28, 2016
   Lembe Lokk, voice
   David Rothenberg, clarinet
   Nightingale

Lembe's beautiful song leaves space for the bird's interjections—friend or foe, fact or dream. It really happened, just like this.

3. While Birds Chant Praises (2:38)
   Landwehr Canal, Kreuzberg, May 10, 2017
   Cymin Samawatie, voice
   David Rothenberg, clarinet
   Lembe Lokk, voice
   Nightingale

Cymin sings her own words:

   Today I give my sorrow free rein
   Drench my pain with your deep tones
   Dearest Kim, please don't stop, please don't stop.

   I want to open all my wounds
   And let the tears flow
   In this moment setting wisdom aside.

4. You've Ruined This Bird For Us (4:08)
   Volkspark Hasenheide, April 23, 2016

Korhan Erel, iPad

Nightingale

You heard it here. Korhan Erel messes with a nightingale by sampling his own song using an app called Samplr and remixing it back to the singing bird. Does this ruin our singer for science? Spend a few years out there listening and only then shall you know.

5. The Nightingales Are Drunk (8:09)

Landwehr Canal, Kreuzberg, May 9, 2017

Lembe Lokk, voice

Korhan Erel, iPad

David Rothenberg, clarinet

Nightingale

The famous words of Hafez inspire the sound of the night.

6. Sharawaji Blues (4:48)

Tuulisaaren Park, Helsinki, May 30, 2016

David Rothenberg, clarinet

Thrush nightingale

In Helsinki the nightingales must contend with a night that never gets dark. They don't love this, because they can easily be seen. So they just keep moving. This, the final night of our northern interspecies musicking, was the moment I was most fed up with the whole enterprise. And it's my favorite duet of the season.

7. Willow Wind (3:27)

Tuulisaaren Park, Helsinki, May 28, 2016

David Rothenberg, *seljefløyte*

Thrush nightingale

Here I play the ancient Norwegian overtone willow flute called the *seljefløyte*, which plays only the pitches of the natural harmonic series. Does the bird get this?

8. No One Sings at Dawn Alone (6:55)
   Tuulisaaren Park, Helsinki, May 28, 2016
   David Rothenberg, bass clarinet
   Thrush nightingale, blackbirds

Approaching dawn, which in Finland in May is truly the middle of the night, the nightingale sounds above the growing chorus of blackbirds and the low thrum of the bass clarinet.

9. The Morning Electric (3:23)
   Tuulisaaren Park, Helsinki, May 30, 2016
   David Rothenberg, iPad
   Thrush nightingale, corn crake, sedge warblers

Dawn brings the machine. Our bird deals with textures, not notes, and his fellow singers find their place in the mix as well.

10. *Sisichak* (4:19)
   Mäntyharju, Finland, May 28, 2016
   David Rothenberg, *furulya*
   Blyth's reed warblers

By chance at dawn in central Finland we stumble into a jam session of Blyth's reed warblers, the most musical of European warblers, a bird that Geoff Sample says could be renamed the *sisichak*, because that is somehow what it sounds like. They riff and play together with sound, seemingly neither defending territories nor trying to impress the girls. I join in on a Bulgarian double whistle called *furulya*.

11. Alien Beauty (5:42)
   Tuulisaaren Park, Helsinki, May 30, 2016
   David Rothenberg, iPad, clarinet
   Thrush nightingale, sedge warblers

Later that last Helsinki morning, a more precise electrified sound emerges in the midst of nature—does it make even a drum beat of sense?

12. She's Finally Here (3:59)
    Volkspark Humboldthain, Berlin, May 9, 2018
    David Rothenberg, clarinet
    Benedicte Maurseth, Hardanger fiddle
    Nightingale

We finally hear the short, attenuated phrases of a male nightingale which happen only once a female has arrived. All that singing finally gets him somewhere . . .

13. I Cannot Go Home (3:59)
    Floraplatz, Tiergarten, Berlin, May 7, 2018
    David Rothenberg, half-clarinet
    Wassim Mukdad, oud
    Volker Lankow, frame drum
    Ines Theileis, voice
    Nightingale

Our bird can sing over a beat. Does he hear the pull of regulated time, or is it all just noise in his way? "So easy for this bird to travel thousands of miles," muses Wassim. "I had to cross so many borders to get here, and should I return home to Syria, I would probably be killed. In my homeland I was a doctor, and an activist against the war. Here in Berlin I am a musician. Sometimes life goes that way."

14. Exit Music (4:04)
    Viktoriapark, Kreuzberg, Berlin, May 5, 2017
    Lembe Lokk, voice
    Sanna Salmenkallio, violin
    David Rothenberg, clarinet
    Police, closing us down
    Nightingale

All right, the final piece, nearly 1:00 A.M., when the neighborhood has had enough. Reminds me of the end of *Monty Python and the Holy Grail*, when the cops show up and haul everyone away.

15. Nightingale, You Are the One (6:56)
   Viktoriapark, Kreuzberg, Berlin, May 5, 2017

Finally our bird can sing for himself, alone, no humans to trouble him. The wild will win in the end.

Total time 75:00
Recorded live by David Rothenberg, Ville Tanttu, and Reelika Ramot
Mixed and mastered by David Rothenberg
All titles published © 2019 Mysterious Mountain Music (BMI)

# NOTES

## Chapter One

1    Cited in Richard Mabey, *The Book of Nightingales* (London: Sinclair-Stevenson, 1997), 30.

2    Oliver Pike, *The Nightingale: Its Story and Song* (London: Arrowsmith, 1932), 20.

3    Ibid., 21.

4    Letter from Luxemburg to Sophie Liebknecht, Wronke, end of May 1917, in Rosa Luxemburg, *Letters from Prison to Sophie Liebknecht*, translated by Eden Paul and Cedar Paul. http://www.marxists.org/archive/luxemburg/1917/undated/01.htm.

5    Richard Prum, "Coevolutionary Aesthetics in Human and Biotic Art-worlds," *Biology and Philosophy* 28, no. 5 (2013): 811–32.

## Chapter Two

1    Marvin Minsky speaking at the conference celebrating the sixtieth birthday of the composer R. Murray Schafer, Banff Centre, Alberta, 1993.

2    Evan Eisenberg, *The Recording Angel* (1987; New Haven, CT: Yale University Press, 2005), 206.

3    John Berger, *Why Look at Animals?* (1977; London: Penguin, 2009).

## Chapter Three

1    Pauline Oliveros, *Sounding the Margins* (Kingston: Deep Listening, 2010).

2   Percy Shelley, *A Defense of Poetry* (1821; Indianapolis: Bobbs-Merrill, 1904), 30. https://archive.org/details/defenceofpoetry012235mbp.

# Chapter Four

1   Roger Payne and Scott McVay, "Songs of Humpback Whales," *Science* 173 (1971): pp. 585–97.

2   T. S. Collett, "Pulling the Wings off Flies," *Nature* 401, no. 6748 (1999): 12.

3   Thomas Nagel, "What Is It Like to Be a Bat?," *Philosophical Review* 83 (1974): 435–50.

4   Michael Weiss, Sarah Kiefer, and Silke Kipper, "Buzzwords in Females' Ears? The Use of Buzz Songs in the Communication of Nightingales (*Luscinia megarhynchos*)," *PLoS ONE* 7, no. 9 (2012). However, others have argued that the whistle songs of male nightingales may be the sexiest syllables: see Hansjoerg P. Kunc, Valentin Amrhein, and Marc Naguib, "Acoustic Features of Song Categories and Their Possible Implications for Communication in the Common Nightingale (*Luscinia megarhynchos*)," *Behaviour* 142, no. 8 (August 2005): 1077–91.

5   Richard Prum, "The Lande-Kirkpatrick Mechanism Is the Null Model of Evolution by Intersexual Selection," *Evolution* 64 (2010): 3085–3100.

6   Olavi Sotavalta, "Song Patterns of Two Sprosser nightingales," *Annals of the Finnish Zoological Society "Vanamo"* 17, no. 4 (1956): 5.

7   Ofer Tchernichovski et al., "Studying the Song Development Process," *Behavioural Neurobiology of Birdsong, Annals of the New York Academy of Sciences* 1016 (2004): 348–63. See also Ofer Tchernichovski, Partha Mitra, et al., "Dynamics of the Vocal Imitation Process: How a Zebra Finch Learns Its Song." *Science* 291 (2001): 2564–69.

8   I have written extensively about this elsewhere in *Why Birds Sing*, the same story with whales in *Thousand Mile Song*, and in greater detail on why it is important to make music live along with animals in David Rothenberg, "Interspecies Improvisation," *Oxford Handbook of Critical Improvisation Studies*, vol. 1, ed. George Lewis and Benjamin Piekut (London: Oxford University Press, 2016), 500–522.

9   Elizabeth Hellmuth Margulis, "Aesthetic Responses to Repetition in Unfamiliar Music," *Empirical Studies of the Arts* 30 (2013): 45–57. See also Elizabeth Hellmuth Margulis, *On Repeat: How Music Plays the Mind* (New York: Oxford University Press, 2013), 15.

10  You can hear Matthew Barley's fine performance with us and a British

nightingale on soundcloud at https://soundcloud.com/terranova/matthew
-barley-live-with-a-nightingale.

11   David Rothenberg, Tina C. Roeske, Henning U. Voss, Marc Naguib, and
Ofer Tchernichovski, "Investigation of Musicality in Birdsong," *Hearing Research* 308 (2014): 71–83.

12   Tina C. Roeske, Damian Kelty-Stephen, and Sebastian Wallot, "Multifractal
Analysis Reveals Music-like Dynamic Structure in Songbird Rhythms," *Scientific Reports* 8, article no. 4570 (2018), https://www.nature.com/articles
/s41598-018-22933-2.

## Chapter Five

1   Almo Farina, Nadia Pieretti, and Rachele Malavasi, "Patterns and Dynamics of (Bird) Soundscapes: A Biosemiotic Interpretation," *Semiotica* 198
(2014): 241–55. See also Almo Farina, *Soundscape Ecology* (Dordrecht:
Springer, 2014).

2   Personal interviews with Fred Jüssi, October 19, 2015, and August 2, 2017.

3   David Quammen with photos by Stephen Wilkes, "How National Parks Tell
Our Story," *National Geographic*, January 2016. http://ngm.nationalgeo
graphic.com/2016/01/national-parks-centennial-text#photographs.

4   Denis Diderot, *Diderot on Art II: The Salon of 1767* (New Haven, CT: Yale
University Press, 1995). http://tems.umn.edu/pdf/Diderot%20On%20Art
%20II.pdf

5   http://hearbirdsagain.com/.

## Chapter Six

1   Hans Slabbekoorn, "Songs of the City: Noise-Dependent Spectral Plasticity
in the Acoustic Phenotype of Urban Birds," *Animal Behaviour* 85 (2013):
1089–99.

2   There is a white parrot named Snowball who is supposed to be the only animal who can follow a beat. I don't believe such behavior is so rare, but these
researchers think it is: Ani Patel, John Iversen, Micah Bregman, and Irena
Schulz, "Experimental Evidencefor Synchronization to a Musical Beat in a
Nonhuman Animal," *Current Biology* 19 (2009): 827–30.

3   Interview with Gordon Hempton by Nika Knight, https://www.guernica
mag.com/interviews/learning-to-listen/.

**4**   John Muir, quoted in Gordon Hempton and John Grossmann, *One Square Inch of Silence* (New York: Free Press, 2009), 245.

**5**   Bernie Krause, personal communication.

**6**   From the last Leonard Cohen interview by David Remnick, http://www .wnyc.org/story/leonard-cohen-last-interview/.

**7**   WBUR's *Stylus* radio show on the Moodus Noises: http://stylusradio.org /post/79188275851/on-a-clear-summer-day-in-the-early-1980s-cathy. See also Brian Kane, *Sound Unseen* (New York: Oxford University Press, 2014).

## Chapter Eight

**1**   See the section "For Further Reading" for a list of books on nightingales, interspecies music, and other relevant sources to go beyond this book.

**2**   Immanuel Kant, *Critique of Judgment*, trans. Werner Pluhar (Indianapolis: Hackett, 1987), 94.

**3**   Geoff Sample, personal communication, March 2018.

## Chapter Nine

**1**   Hafez, trans. Dick Davis, *The Nightingales Are Drunk* (London: Penguin, 2015), 35.

**2**   Peter Handke, *The Moravian Night*, trans. Krishna Winston (New York: Farrar, Straus & Giroux, 2016), 97–98.

# INDEX

Page numbers in italics refer to figures.